GUIDE TO ELECTRICAL INSTALLATION AND REPAIR

McGraw-Hill Paperbacks
Home Improvement Series

Guide to Plumbing
Guide to Electrical Installation and Repair
Guide to Roof and Gutter Installation and Repair
Guide to Wallpaper and Paint
Guide to Paneling and Wallboard

GUIDE TO ELECTRICAL INSTALLATION AND REPAIR

McGRAW-HILL BOOK COMPANY

New York   St. Louis   San Francisco   Auckland   Bogotá   Düsseldorf
Johannesburg   London   Madrid   Mexico   Montreal   New Delhi   Panama
Paris   São Paulo   Singapore   Sydney   Tokyo   Toronto

1 2 3 4 5 6 7 8 9 0 SMSM 8 3 2 1 0

Library of Congress Cataloging in Publication Data

Main entry under title:

Guide to electrical installation and repair.

(McGraw-Hill paperbacks home improvement series)
Originally issued in 1975 by the Automotive-Hardware Trades
Division of the Minnesota Mining and Manufacturing Company
under title: The home pro electrical installation and repair guide.
1. Electric engineering — Popular works. I. Minnesota Mining
and Manufacturing Company. Automotive-Hardware Trades
Division. The home pro electrical installation and repair guide.
TK148.M48   1980   621.319′24   79-14720

ISBN 0-07-045964-9

Front cover photo by Lizabeth Corlett

Contents

IMPORTANT NOTICE—
READ BEFORE
PROCEEDING

The procedures in this book are based on information which we believe to be reliable. By carefully following these procedures and obeying the rules for safety on Pages 19 and 20 when working with electricity, you should be able to accomplish these repair and wiring jobs without hazard to yourself or danger to your home or property.

However, these procedures assume that your present wiring is properly installed in accordance with local and National codes. In many cases, this might not be so. For example, if your home is more than 15 years old the wiring is probably not adequate to provide for the electrical demands of our many modern appliances. Your entire electrical system may need updating – a job best left to the professionals.

Also, if your home has been previously occupied by others, it is possible that the electrical system has been modified. For example, the fuses may have been replaced with improper, higher amperage fuses, thus depriving circuits of needed protection.

Previous parties may also have installed wiring, appliances and circuit protection incorrectly. If, as a result, an appliance motor is not correctly grounded, you could very possibly be severely shocked when you attempt to work on it.

These possibilities suggest one, overriding safety rule which should never be forgotten:

Before undertaking any electrical work, be sure that the existing wiring is correct.

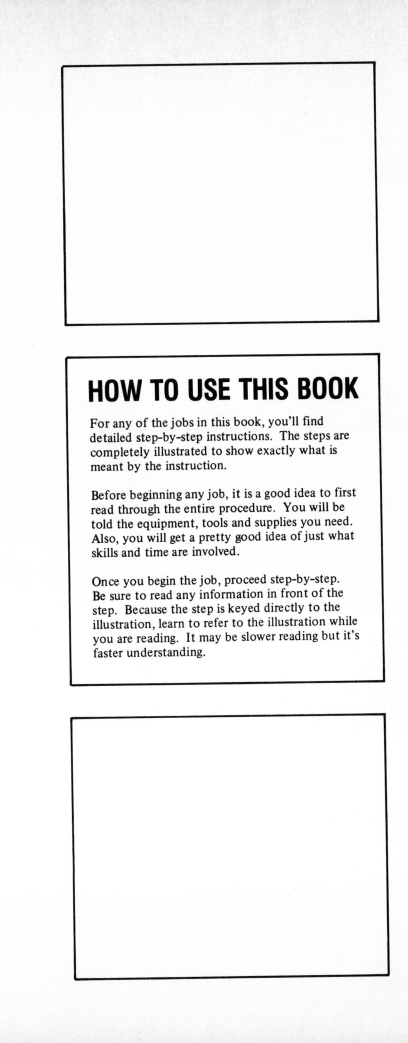

HOW TO USE THIS BOOK

For any of the jobs in this book, you'll find detailed step-by-step instructions. The steps are completely illustrated to show exactly what is meant by the instruction.

Before beginning any job, it is a good idea to first read through the entire procedure. You will be told the equipment, tools and supplies you need. Also, you will get a pretty good idea of just what skills and time are involved.

Once you begin the job, proceed step-by-step. Be sure to read any information in front of the step. Because the step is keyed directly to the illustration, learn to refer to the illustration while you are reading. It may be slower reading but it's faster understanding.

GUIDE TO ELECTRICAL INSTALLATION AND REPAIR

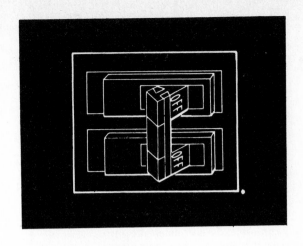

ELECTRICAL SYSTEM

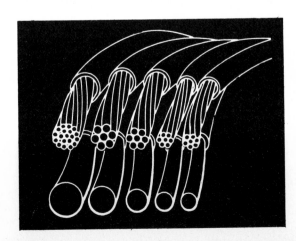

In order to maintain and repair electrical fixtures and wiring in your home, you should become familiar with the parts of the electrical system and what they do.

This knowledge will help you determine which electrical jobs you can accomplish safely and which jobs are best left to the professionals. Because of the potential for hazard to you and damage to your home, do not undertake any electrical repair that you do not understand thoroughly.

This section will explain where electricity comes from, where it goes and in general what it does. It lists the precautions that you should observe.

In most communities, the local power company supplies and maintains the wiring up to your electric meter. Beyond the meter, it becomes your responsibility as part of your home.

Your first step before making any installations, additions, and some repairs is to check with your local city or county building inspector concerning local codes and the National Electrical Code.

You are limited, not only by the National Electrical Code, but also by local (city or county) code. The main goal of both codes is safety, not to prohibit you from doing your own electrical work.

With some exceptions, such as replacing a switch or light fixture, you are required by code to get a permit to do most electrical work. After completion, your work must be approved by an authorized inspector.

Codes not only determine what you may or may not do, but also limit you as to materials that may be used, such as size and type of wire.

Therefore, it is recommended that you contact your building inspector before you attempt any electrical work, except minor repair or replacement of fixtures.

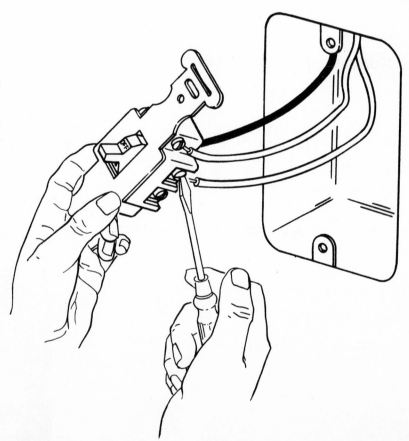

Electricity is produced by power generating plants and then transmitted to distribution stations. From these stations, it is transmitted to local areas through overhead or underground cables. There it branches off into the wires leading to individual homes or buildings.

There are two basic types of home electrical wiring systems. Depending on age of the house and the local requirements at the time of construction, a two-wire or three-wire system is found.

However, before the two types of systems are described, you should be made familiar with common electrical terms and their definitions. This page and the next define common electrical terms.

POWER GENERATING PLANT

DISTRIBUTION STATION

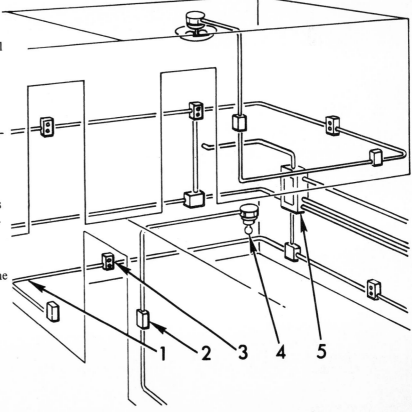

Amp (ampere) — Rate of flow of electrons (small electrically charged particles), or quantity of electrical current.

Circuit — Wires leading from a protection device (fuse or circuit breaker [5]) throughout the house. Each continuous wire from the protection device is called a branch circuit, regardless of number of fixtures [4] or junctions [1].

Fixture — Any mounted electrical device such as a switch [2], outlet [3], ceiling light [4], etc.

Ground — A connection between an electrical wiring system, or part of it, and the earth. Also, the neutral side of a circuit leading to the above connection.

Main Power Panel — A cabinet housing circuit protection devices — either fuses or circuit breakers.

3

TYPES OF SYSTEMS

Neutral — The white wire [7] in a service entrance. It is connected to the common ground connection [4] in the fuse or circuit breaker panel [3]. Also, all white wires [1] throughout the house are neutral.

Overload — When a wire, switch or other fixture in an electrical circuit carries more current than it can safely handle, the condition is called an overload. An overload is often caused by too many appliances on one circuit or by a faulty appliance motor.

Service Entrance — The cable or lines, usually in conduit [5], which connect with the power company's lines on the outside of the house. Also included in the service entrance are the electric meter [6], main switch [2], fuse or circuit breaker panel [3] and ground connection [4].

Short — When a "hot" or "live" wire [2] touches a neutral or grounded wire [1] or component due to worn insulation, loose connections or numerous other reasons, a path to ground is created. The result is known as a shorted circuit or "short".

Volt — The pressure that starts electric current and keeps it moving — more commonly called voltage. Household voltage may be any value from 110 to 120 volts and 220 to 240 volts. For simplification, the value of household voltage, as used in this book, will be 115 and 230 volts unless otherwise specified.

Watt — A measurement of total electrical energy (power) flowing in a circuit at any given moment.

Watt-hour — Watts consumed during one hour.

Kilowatt-hour (kwh) — 1,000 watt-hours. You pay your electrical bill according to the amount of kilowatt-hours consumed during a specified length of time such as one or two months.

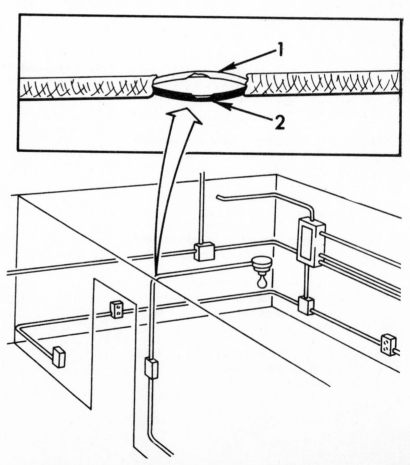

▶ Two-Wire Systems

Most smaller homes built during the mid-forties and earlier have the basic two-wire power system.

Only two wires [1] enter the house from either a pole or underground cable. These wires bring 115 volts to the meter [2]. They continue to the main power panel [3] and to the fuse box [5], where individual circuit fuses [4] are contained.

The white or gray wire is called the ground wire. It is usually connected to the nearest cold water pipe. The black wire is called the "hot" or "live" wire. Both wires actually have current.

The two-wire system in many cases does not provide adequate power for television sets, power tools and numerous other electrical appliances used in the home today. The two-wire system usually provides only 30 amp 3,600 watt power to your home.

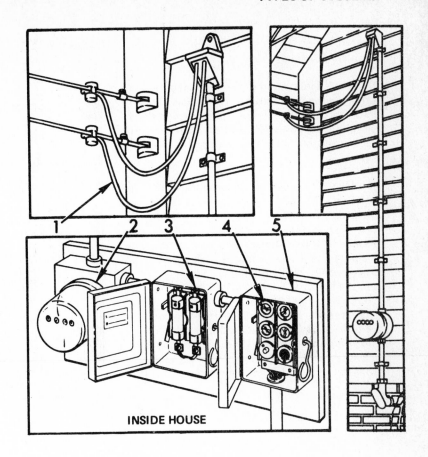

INSIDE HOUSE

▶ Three-Wire Systems

The three-wire system is used in most homes today. This system provides both 115 and 230 volts, offering full use of minor and major electrical appliances.

The three service entrance wires [1] terminate at the meter [2]. From the meter, they go to the fuse or circuit breaker panel [3] and then they branch into individual circuits.

There is one white wire, called the neutral wire, and two black wires. (Usually the "hot" wires are black. However, they could be some other color, but they are never white or green.)

One hundred fifteen volts is available from between a black wire and the neutral wire. Thus, there are two 115 volt branches to a house. They are distributed throughout the house in a way that balances the amount of power which must be provided by either branch. The two black (115 volt) wires are joined inside the power panel to also provide your home with the 230 volts required by some appliances such as ranges and dryers.

It should be noted that the actual voltages most commonly used in homes today are 120 and 240 volts. Although 110 and 220 volt current was once common in this country, it is no longer used. For purposes of making electrical calculations, 1971 National Electrical Code requires that all determinations be based on 115 and 230 volts.

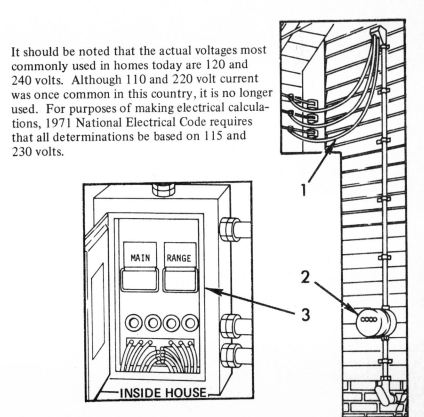

MAIN   RANGE

INSIDE HOUSE

Three-Wire Systems

The three-wire system provides 60 amp 14,400 watts. In the newer homes, it provides 100 amp 24,000 watts, which is quite adequate for most residential purposes. Central air conditioning or electric house heating may require even more power.

In kitchens with an electric range, the three-wire system must be used with a single, fused 230 volt circuit [1] for the range.

Because of the existence of two different voltages in a home, precautions must be taken so that the lower voltage (115 volts) appliance is not plugged into the higher voltage source.

To prevent such an error, different type outlets are provided for 115 volts and 230 volts. The 230 volt outlets are usually the "crow-foot" [3] or L-shaped type [2]. This design prevents you from inserting a 115 volt plug into a 230 volt outlet.

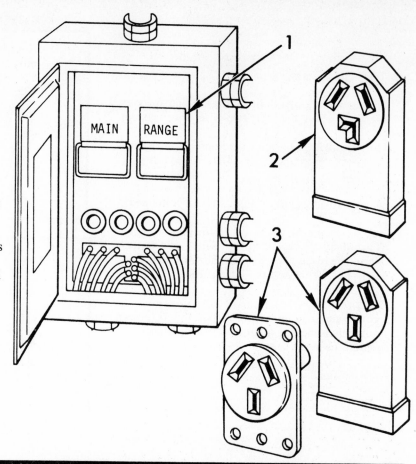

CIRCUITS

The wires leading to various parts of your house begin just in back of the protection device (fuse or circuit breaker box). A circuit is made up of all wiring that is controlled by the same fuse or circuit breaker.

There are three different types of circuits in most homes:

- General purpose circuits, see next section (below).
- Special appliance circuits, Page 7.
- Major appliance circuits, Page 7.

▶ General Purpose Circuits

General purpose circuits are used for all lighting and for the outlets in all rooms except the kitchen and possibly the workshop.

These circuits are 115 volts. They are generally wired with Number 14 wire and protected by a 15 amp fuse or circuit breaker. Each circuit has a maximum capacity of about 1,750 watts.

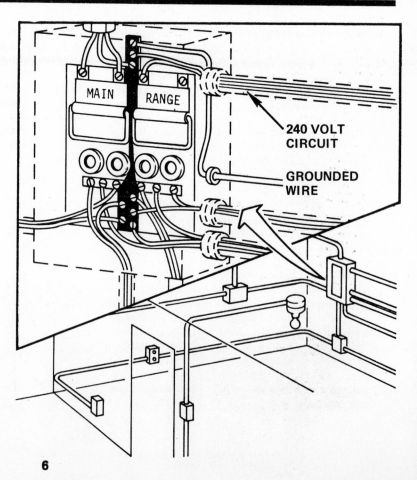

240 VOLT CIRCUIT

GROUNDED WIRE

General Purpose Circuits

Circuits for convenience outlets and lighting are usually planned so that there is one circuit for every 350 to 500 square feet of floor space.

If a general purpose circuit is wired with Number 12 wire and protected by a 20 amp device, it has a capacity of 2,300 watts.

Here, also, codes will determine how your general purpose circuit is wired.

▶ Special Appliance Circuits

Special appliance circuits are installed to provide power to rooms such as the kitchen, breakfast room, laundry and workshop.

These 115 volt circuits are usually wired with Number 12 wire and protected by a 20 amp device.

They provide power for small kitchen appliances such as portable broilers, food blenders and toasters.

It is not necessary to have one circuit for each appliance, since rarely are more than two appliances used at one time.

▶ Major Appliance Circuits

This type of circuit provides 230 volts to an individual circuit for each high-wattage appliance, such as an electric range, dryer or water heater.

The size of the wire and the current rating of the protection device depend on the wattage rating of the major appliance.

WIRES AND CABLES

▶ Wires

Wires are used to carry electricity throughout your house from the service entrance to any point where it is used. The size of these wires depends upon local codes and National Codes and their end use such as lighting, heating or air conditioning.

Copper wires are used in almost all home wiring. However, more and more aluminum wire is being made due to the shortage of copper.

Wires carrying electrical current can be compared to pipes carrying water — the larger the pipe (wire), the more water (power) it can carry. Therefore, the more current (amps) required, the larger the wire must be.

Wire sizes are based upon the American Wire Gage (AWG) system. The diameter of the wire determines its gage. The larger the diameter, the smaller the gage. The sizes of wires are expressed in terms of Numbers or Gages, i.e., Number 14 wire is 14 gage wire.

Wire gage is directly related to wire diameter. It makes no difference whether a wire is made up of a single large strand [1] or of many small strands [2] such as used in appliance cords. It is the total diameter of the wire that is important.

The trend is to make the larger wires multi-strand rather than a single large strand. The large strands are too stiff to handle easily. For wire sizes Number 8 and larger (6, 4, 2, 0), multi-strand construction is required by Code.

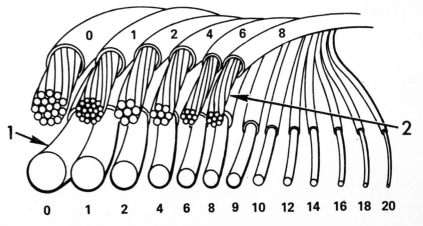

WIRES AND CABLES

Wires

Common sizes for home wiring range from Number 0 to Number 18. Most home power lines range from Number 10 down to Number 14. However, some heavier circuits require larger wires. Number 16 and 18 wires are used in the home for door bell or intercom wiring.

Wires are insulated with a variety of rubber and plastic covering. The wire size [1] is usually printed on the insulation itself. For home wiring the more common types that are available and their use are listed in the chart at right.

Type	Material	Use
R	Rubber	General (this type may be in use, but is no longer being made)
RH	Heat-resistant rubber	General
RU	Latex rubber	General
RW	Moisture-resistant rubber	General and wet locations
RH-RW (RHW)	Moisture-and heat-resistant rubber	General and wet locations
T	Thermoplastic	General
TW	Moisture-resistant thermoplastic	General and wet locations
THW	Moisture-and heat-resistant thermo-plastic	General and wet locations

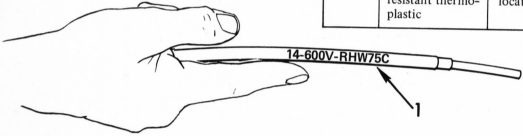

▶ Cables

Often it is desirable to group 2 or more wires in the form of a cable. This makes it easier to route the wires through wall and ceiling spaces.

Cables are classified by the size and number of wires they contain. For example, if a cable has three Number 14 wires, it is termed "14-3" cable. If it also contains a bare ground wire, it is known as "14-3 with ground".

The four more common cables are described in the following paragraphs.

▶ Armored BX Cable

Armored BX cable is used only in dry, indoor locations. The spiral armor [1] is made of galvanized steel.

Around the wires is a spirally wrapped layer of insulating paper [2] to protect the wires from the armor.

Between the paper and the armor is a bare grounding wire [3]. BX cable [1] is made in 2- and 3-wire types plus ground wire. This type of cable must be used with steel electrical boxes [4] only.

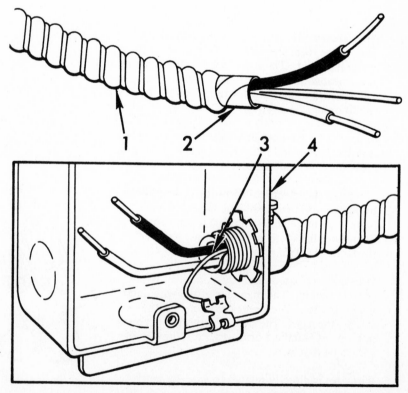

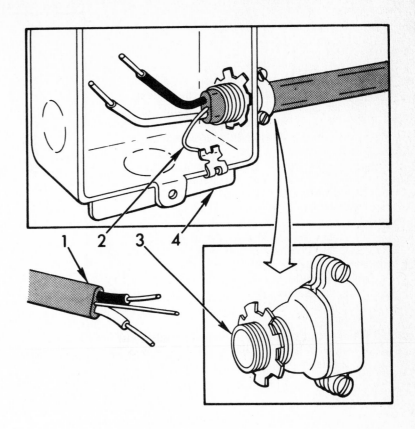

▶ Nonmetallic Sheathed Cable

This type of cable is used only indoors. The outer covering [1] is flame and moisture resistant. It is made either with or without a grounding wire [2].

Nonmetallic sheathed cable is becoming more popular because it is easy to install, lightweight and comparatively less expensive than other types.

Nonmetallic cable uses a flat-shaped connector [3] when connecting to electrical boxes [4].

▶ Conduit

Conduit is not a cable. It is a pipe made to protect cables. There are three basic types of conduit: rigid, thin-wall, and flexible conduit.

Rigid conduit [1] resembles water pipe. The inside surface of conduit differs from water pipe in that it is carefully finished and prepared so that wires can easily be pulled through without damage to their insulation.

Rigid conduit is usually made in 10-foot lengths. A conduit bender [2] is used to bend the pipe for turning corners and angles.

The conduit is installed first. The wires or cables [3] are then pulled through the conduit. The conduit is threaded at the ends and uses conduit fittings [4] similar to BX cable fittings.

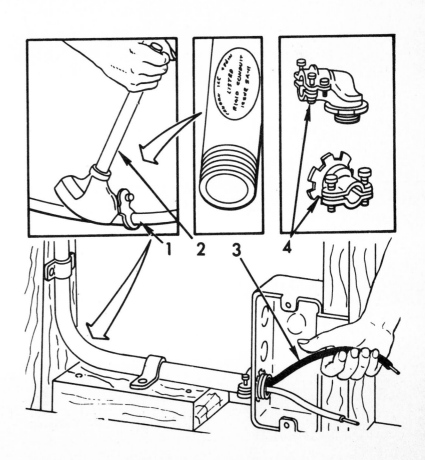

WIRES AND CABLES

Conduit

Thin-wall conduit [1] is similar to rigid conduit except that it is not threaded. The walls are thinner, as the name implies, than rigid conduit. For the same size conduit the inside diameter of thin-wall conduit is slightly larger than rigid.

Thin-wall conduit fittings and connectors are not threaded. When a nut is tightened, pressure is exerted on a slip ring which holds the conduit tightly.

Flexible conduit [3] is similar to armored BX cable, but does not contain wires. It is also larger in size. Flexible conduit is installed first and the wires are pulled through afterward. Armored BX cable connectors [2] are used for joining flexible conduit to fixtures.

This type of conduit is seldom used today, but whenever it is used, it is used just as armored BX cable is used.

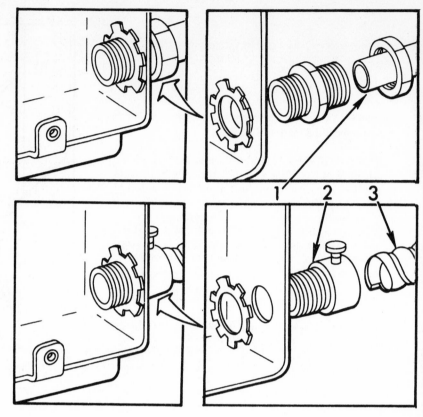

▶ Underground Cable

There are several types of underground cable [1]. Depending upon your local codes, you may be required to run the cable through buried conduit [2] or you may be allowed to simply install a ready-made underground cable without additional protection.

The cable [1] is constructed so that it can be used underground by itself. The insulated wires are encased in a plastic material that is moisture resistant and is tough enough so that it will not be damaged if accidentally hit with a shovel.

Underground cable must never be spliced. It must be installed in a continuous run from power source to fixture.

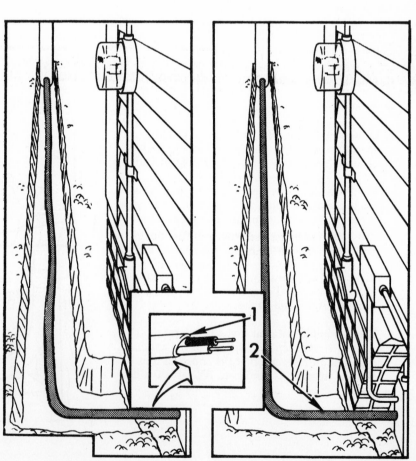

A protective device such as a master switch or circuit breaker must be installed at or close to the electrical service entrance. It permits all power to be disconnected from the entire house. All power must be disconnected when making some types of major repairs, or in case of an emergency such as a fire.

Usually the main disconnect and branch circuit protection devices are housed in a single metal cabinet [1, 2]. Sometimes, however, main disconnects may be located in a separate box.

All main power switches must be operated externally. That is, a person must be able to turn the main power off without being exposed to "hot" or "live" wiring.

There are three basic types of main power disconnects:

    Drawer type [1]
    Circuit breaker [2]
    Lever type [3]

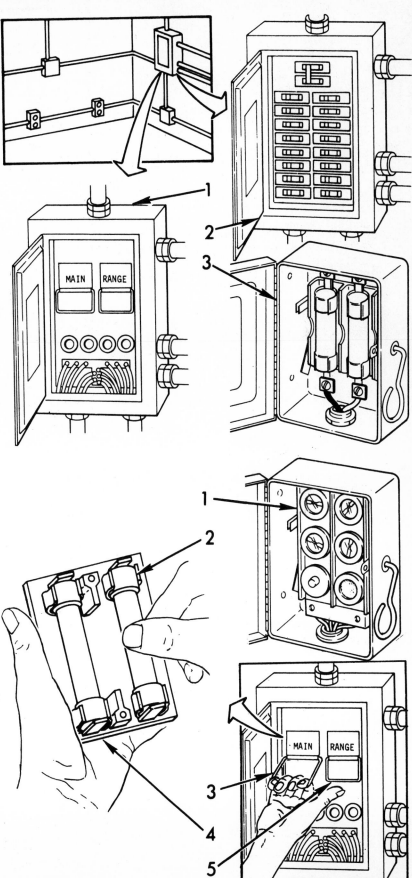

In most older homes, the main power disconnect has the lever type of disconnect [1]. By pulling an external handle you will shut off power to the entire house. The door to the fuses cannot be opened until the handle has been placed in the OFF position. Therefore, before you can change a fuse, all power to the house must be disconnected.

Most main switches used today in fuse boxes are not the lever type. Instead, the main cartridge fuses [2] are mounted in an insulated drawer [4] that can be pulled out of the box. This action results in power to all of the circuits being shut off.

The drawers may all be for main power shut-off [3], or one may be for an electric range [5] or similar high-current-rated appliance.

MAIN POWER PANELS

Once the drawer(s) [1] have been removed, all circuits are disconnected. You should always remove the drawer before changing a branch circuit fuse [2]. To connect power again, simply push the drawer into the opening. The drawer will not connect power if it is inserted upside down.

The newer type circuit breaker panels [3] have one or two large switches [4, 5] under the cover or door of the box. To shut off power, simply push the switches to the OFF position.

There are two methods of turning power back on, depending on make of circuit breaker.

● Simply push switch [4, 5] to ON position. Power is restored.

    OR

● Push switch [6] past the OFF position (to the RESET position on some makes), then back to the ON position. Power is restored.

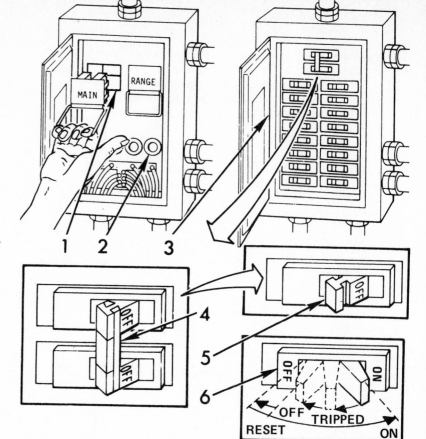

▆▆ FUSES ▆▆▆▆▆▆▆▆▆▆

A fuse [1] is a protective device with a soft strip of metal [2] through which current flows. This strip of metal is designed to melt whenever too much current flows through it. Possible causes are:

● Too many appliances are plugged into one circuit (an overload).

● A "hot" or "live" wire touches a neutral wire or is grounded through a fixture or appliance (short circuit).

When either an overload or short circuit occurs, the power to that circuit is interrupted and the fuse is said to "blow".

There are several different types of fuses used as protective devices in the home. The more common types will be discussed in this section.

Plug-type fuses [3], Page 13.

Non-tamperable fuses [4], Page 13.

Screw-in breakers [5], Page 14.

Time-delay fuses [6], Page 14.

Cartridge fuses [7], Page 14.

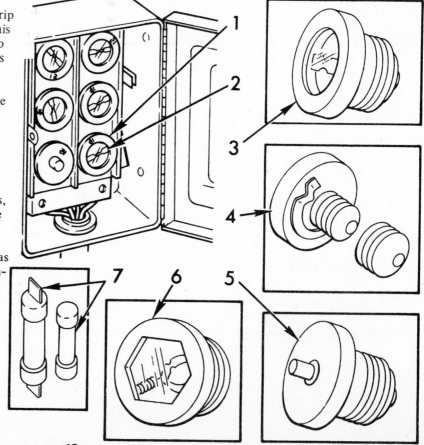

▶ **Plug-Type Fuses**

Plug-type fuses [1] have a light bulb socket base [2]. They are sometimes called Edison base fuses. They are replaced just as you would replace a light bulb.

This type of fuse was standard and used in most homes until the last several years when newer and improved type fuses have come on the market. In fact, the Code prohibits the installation of these fuses in new construction. This change was made mainly because the basic plug-type fuse is easily interchangeable with one of a higher rating, thus removing adequate protection from the circuit.

When the wiring in a circuit becomes too hot due to an overload or short, the strip of metal will fail. In cases of an overload, the strip of metal will melt at its weakest point, breaking that circuit [3].

In case of a short, the window will blacken [4]. The metal strip will also melt, but you probably won't be able to see it.

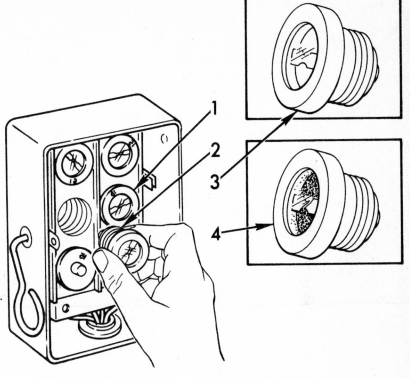

▶ **Non-tamperable Fuse (Type S)**

To prevent "overfusing", or installing a larger size (rating) fuse than the size for which the circuit was designed, non-tamperable fuses [1] were designed. This type of fuse is called Type S by the National Electrical Code.

The non-tamperable fuse consists of two parts — an adapter [3] and the fuse [1] itself. The adapters are rated the same value as the fuses. For example, a 15 amp fuse has a 15 amp adapter. A 20 amp fuse will not fit into the 15 amp adapter. This feature prevents someone from substituting a larger size fuse and "overfusing" the circuit.

These adapters fit into regular plug-type fuse sockets, and once installed, cannot be removed.

Whenever you install a non-tamperable fuse, be sure that you turn it as far as possible — even after it seems to be tight.

There is a spring [2] under the shoulder of the fuse. If this spring is not flattened, the fuse will not make full contact with the socket. As a result, current cannot flow through and you will have no power in the circuit.

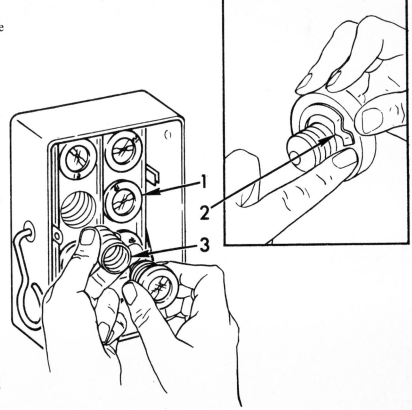

FUSES

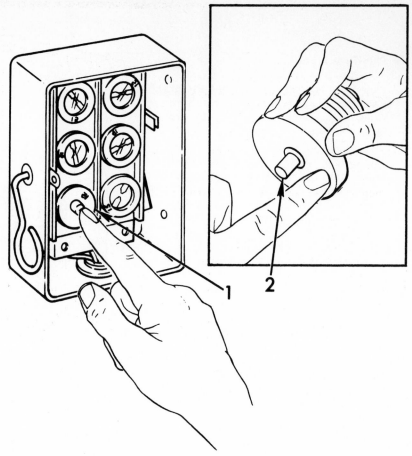

▶ Screw-in Breaker

The screw-in breaker [1] was designed to replace the plug-type fuse in regular fuse panels without changing the wiring.

This type of breaker is useful wherever an occasional overload occurs — such as in a workshop.

Whenever an overload or short occurs, the button [2] pops out. After correcting the problem that created the overload or short, simply push in the button to restore power to the circuit.

▶ Time-Delay Fuses

A time-delay fuse [2] is like a plug-type fuse except that it has a spring-loaded metal strip inside. This feature allows a circuit breaker to have a temporary overload without interrupting the power. The temporary overload may be the extra power required to start such appliances as washers, refrigerators, or power tools. They require much more power to start than to operate. This starting power may be greater than the rating of the circuit. It would repeatedly "blow" conventional plug-type fuses even though the circuit itself was not in danger of being overloaded beyond safety limits.

In case of a short, however, the fuse will blow just as quickly as the plug-type fuse, thus protecting your wiring. For these reasons, when a regular plug-type fuse fails, you should replace it with a time-delay fuse of the same size.

▶ Cartridge Fuses

Cartridge fuses [1] are used mostly for main power fusing in homes. They are usually installed in pull-out drawers so that they may be removed safely. In some older homes, the individual branch circuits are also protected by cartridge fuses.

Cartridge fuses are made in sizes from 15 to 60 amps. The fuse has metal end caps. THE END CAP ON THE METER SIDE OF THE FUSE IS ALWAYS HOT, even if the fuse is blown. See Page 23 for the safest way to remove cartridge fuses.

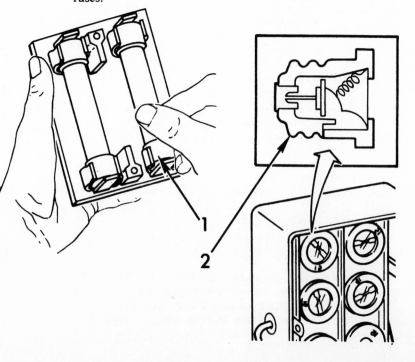

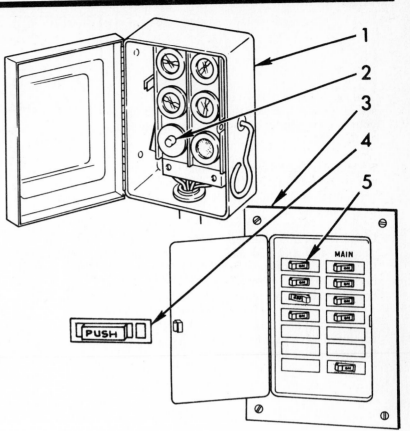

Many homes built since the late 1940's and most homes built today have a circuit breaker panel [3] instead of a fuse panel [1].

Circuit breakers [2, 4, 5] are rated in amperes, just as fuses are. They are available in the same sizes as fuses.

There are three common types of circuit breakers:

Screw-in breaker [2], Page 14.

Switch type [5], below.

Push-button type [4], below.

The switch type [5] is by far the most common circuit breaker.

▶ **Switch Type**

Switch-type circuit breakers [5] used in the home look something like a large switch. They appear similar to the light switches on the wall.

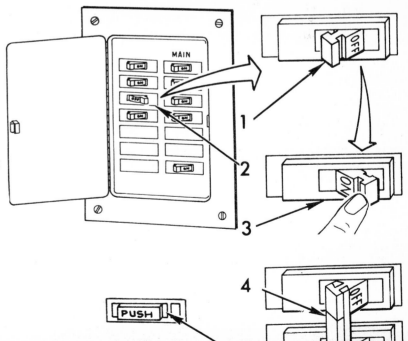

Switch Type

A switch-type circuit breaker does basically the same job as the wall switch. It opens a circuit. This action can be done either automatically or manually.

When an overload or short occurs, the circuit breaker [1] automatically opens the circuit or "trips". After the problem has been corrected, it is simply placed back to ON [3] or, if required, to RESET and then to ON.

The main circuit breaker and the individual circuit breakers for high-wattage-rated appliances are physically larger in size than the branch circuit breakers. Often they are made up of two circuit breakers [4] joined by a connecting bar.

▶ **Push-Button Type**

The push-button type circuit breaker [5] works on the same principle as the switch type [2]. The only difference is that instead of a toggle, a push button is used.

When an overload or short occurs, the button will pop out. It will then stick out farther than the other buttons. After correcting the problem that caused it to pop out, simply push it in to restore power.

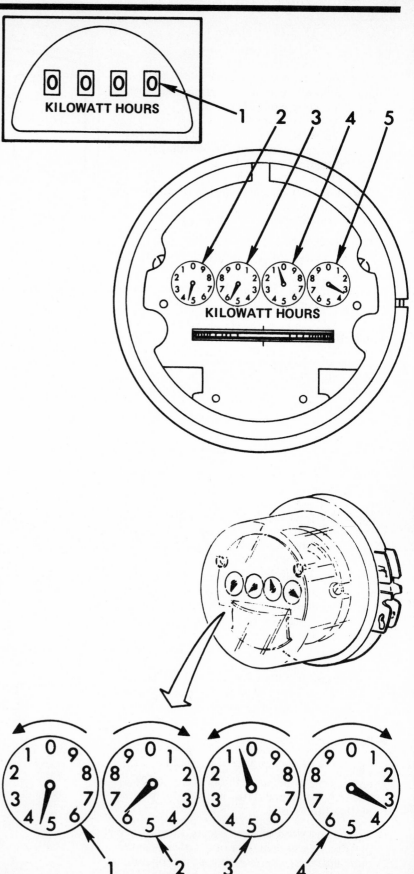

You may want to read your electric meter for several reasons — among them being able to figure your monthly utility bill or periodic consumption of electricity.

Some meters have a cyclometer dial [1] and are simple to read. Just read the numbers in the small windows from left to right to determine number of kilowatt-hours.

Other meters are a little more difficult to read. They are also read from left to right. On two of the dials [3, 5], the pointer rotates in a clockwise direction. On the other two dials [2, 4], the pointer rotates in a counterclockwise direction.

To determine how many kilowatt-hours have been consumed since your last billing period, follow the instructions starting below.

1. Determine the reading of meter at the beginning of billing period — as shown on utility bill.

2. Write down the last number the pointer has passed on each dial. Be sure to remember that two pointers rotate clockwise [2, 4] and the other two counterclockwise [1, 3].

EXAMPLE

The pointer on the first dial [1] is between the 4 and 5. So write down "4".

The pointer on the second dial [2] points at the 6. Look at the dial immediately to the right. If that pointer has passed the 0, write down "6". If it has not passed the 0, write down "5". In this example, write down "6".

The pointer on the third dial [3] is between 0 and 1 (has just passed 0). So write down "0".

The pointer on the fourth dial [4] has just passed the 3. So write down "3". The number from the meter is 4603.

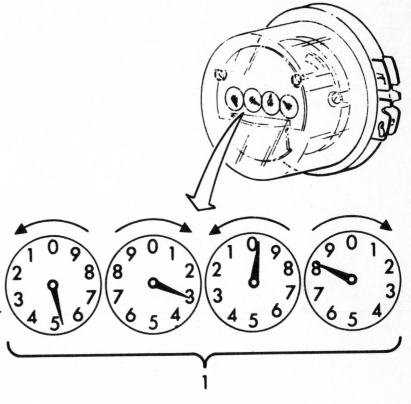

3. After any given period of time, take another reading from the meter dials [1].

4. Subtract the reading obtained in Step 2 from the reading obtained in Step 3.

The result is the number of kilowatt-hours consumed between the two dates.

EXAMPLE

May 15 (Step 3)	5298
March 15 (Step 2)	-4603
Total kwh consumed	695

With this number (695) and the cost per kilowatt-hour, found by calling your local utility company, you can determine the amount of your electric bill.

1

WATTAGE RATING CHART

It is a good idea for you to know the average wattage rating of common devices used in the home. It will help you to determine:

- Whether your wiring is adequate for a branch circuit to accommodate a new appliance.
- The approximate cost per hour to operate an appliance.

To determine whether you can add another appliance to a circuit:

From the Chart, Pages 18 and 19, find the wattage rating of each appliance on the circuit. Don't forget the lights. Add them to determine the total wattage. Call this total "A".

Multiply the voltage (shown on a nameplate on your electric meter) by the size of the fuse or circuit breaker for that circuit. Call this amount "B".

Compare figures "A" and "B". If figure "A" is larger than "B", the wiring in that circuit is not adequate to handle any more appliances. One or more appliances may have to be moved to another circuit to accommodate a new appliance.

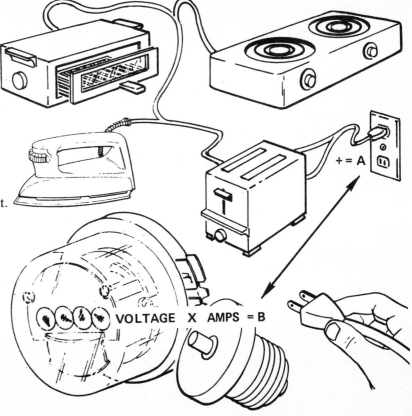

+ = A

VOLTAGE X AMPS = B

WATTAGE RATING CHART

To determine the cost per hour to run an appliance:

Find the wattage rating of the appliance from the Chart, below, or from the nameplate on the appliance. Multiply the wattage rating by the cost per kilowatt-hour of electricity. You will have to call your local utility company to determine the cost per kilowatt-hour. Now move the decimal point three places to the left.

EXAMPLE

Assume that the appliance is a washing machine rated at 500 watts.

Assume that the cost of electricity in your community is 4 cents per kilowatt-hour.

Washing machine	500
Cost of electricity	.04
	20.00

Now move decimal point three places to the left
.02

The cost per hour to operate a washing machine rated at 500 watts is 2 cents per hour.

The Wattage Rating Chart gives the approximate rating of an appliance. The rating of an appliance will differ with different manufacturers. There is usually a wattage rating stamped on the appliance's nameplate.

		Approximate Rating			Approximate Rating
Air Conditioner	(central)	5000	Fryer, deep fat		1400
	(room, 1/3-ton)	800	Frying pan, automatic		1100
	(room, 3/4-ton)	1300	Furnace, coal		400 (see nameplate)
	(room, 1-ton)	1600	gas		150 (see nameplate)
Blender		250 – 1000 (see nameplate)	oil		750 (see nameplate)
Bottle warmer		400	Garbage disposer		400 – 900 (see nameplate)
Broiler		1500	Grill		1000
Can opener		150	Hair dryer		260
Clock		2	Heater (hot water)		2000 – 5000 (see nameplate)
Coffee maker		500 – 1000 (see nameplate)	(room)		1250
Corn popper		500	Hot plate (per burner)		750
Dishwasher		1000 – 1500 (see nameplate)	Iron (hand)		1050
Dryer		4000 – 8000 (see nameplate)	(mangle)		1600
Electric blanket		200	Knife sharpener		100
Fan (attic)		400	Lamp (heat)		250
(exhaust for range)		250	(sun)		400
(portable)		100	Lights (fluorescent circlines)		22 – 32
Floor polisher		350	(fluorescent tubes)		15 – 60
Food warmer		500	(incandescent, per bulb)		10 and above
Freezer, frostless		350 – 500 (see nameplate)	(night light)		7
standard		250 – 400 (see nameplate)	Microwave oven		600 (see nameplate)
(The larger the freezer, the higher the rating)			Mixer		150

Continued on next page

	Approximate Rating		Approximate Rating
Power tools	(see nameplates)	Roaster	1350
Drill, 1/4-inch	150	Rotisserie	1400
3/8-inch	250	Sewing machine	75
1/2-inch	350	Shaver	10
Grinder	200	Stereo hi-fi	300
Lathe	300	Sump pump	300
Lawn mower	300	Television, black and white	250
Sander, portable	750	color	300
Saw, band	250	Toaster	1100
bench	300 – 600	Vacuum cleaner	300 – 800 (see nameplate)
jig	250	Waffle iron	900
radial arm	1500	Washing machine	600 (see nameplate)
sabre	200		
Soldering iron	150		
Projector (slide or movie)	350		
Radio, console	150		
portable	50		
Range, oven	4000 – 8000 (see nameplate)		
top	4000 – 5000 (see nameplate)		
Refrigerator, frostless	300 – 450 (see nameplate)		
standard	250 – 350 (see nameplate)		
(The larger the refrigerator, the higher the rating)			

SAFETY WITH ELECTRICITY

Working with electricity is not dangerous as long as you strictly obey safety rules and treat electricity as a potential danger.

Although there are many rules to follow, they are basically the same: Be careful — don't work on electrically "hot" fixtures or appliances. The safety rules given in this section should always be followed to prevent damage, injury or worse.

The three most important rules that will protect you and your home are:

1. PLAY IT SAFE — Be sure the main power, or at least power to the current being worked on, is off.

2. If you are not knowledgeable about the electrical project being undertaken — don't do it. Do not take on an electrical project that is too complicated. Let a professional handle it.

3. Always check local codes before attempting your own wiring.

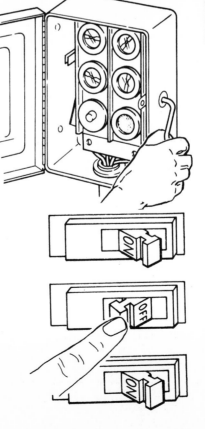

SAFETY WITH ELECTRICITY

Follow these additional rules for safety:

4. Know where the main power shutoff is located. Be sure you can reach it and turn it off in case of an emergency.

5. Always replace fuses with the correct size. Keep extra fuses in the fuse cabinet. Store at least one fuse for each circuit.

6. Never touch a bare wire or any metal that is touching a bare wire.

7. Do not touch electrical fixtures, switches or appliances when you are wet or standing on wet ground.

8. Do not plug too many appliances or extension cords into one outlet.

9. Any fixture, appliance or cord that gives a shock or spark should be repaired or replaced immediately. Do not wait for something more serious to happen.

10. After using appliances, especially heating appliances, always turn them off before leaving them. Turn them off even if you are leaving only for a short time.

11. If combustible materials, such as paper or cloth, are used as lamp shades, be sure they are not in contact with the bulb.

12. Warn children not to climb utility poles or trees near electrical wires. Warn them against flying kites near utility poles.

13. Use fixtures and equipment with the "Underwriters Laboratory" seal on them. The Underwriters Laboratory (UL) tests equipment and fixtures for safety. They then "list" this equipment. This means it has passed minimum requirements for safety.

14. If you see anything wrong with electrical lines or poles, call your local power company at once. Do not try to fix it. They will do it.

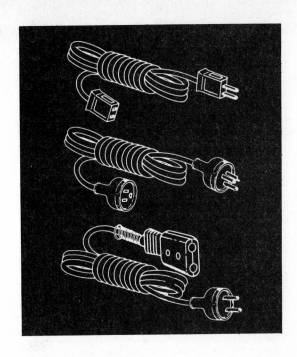

ELECTRICAL REPAIRS

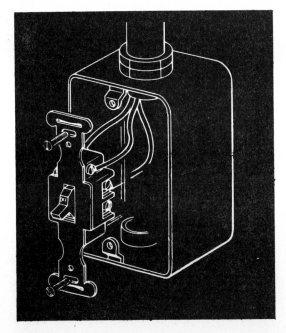

There are many electrical jobs you can do safely and well without being a professional. They include minor repairs such as replacing cords and plugs and repairing or replacing electrical fixtures. Other electrical work such as adding lighting to your garage is more complicated and may require compliance with local and national electrical codes.

First go to Page 19 and read the safety rules. Be sure you know what you are doing before attempting any repair. You should read the entire repair procedure before beginning. This will familiarize you with the problem and tell you what tools and parts you may need.

You probably already have most of the tools required to make common electrical repairs. Some of the more complicated jobs can be accomplished with inexpensive special tools that may be required.

The first part of this section will aid you in identifying the circuit and isolating the problem. Once the problem is identified, go to the appropriate section and follow the procedures to fix your specific problem.

IDENTIFICATION OF CIRCUITS

You should know which fuse or circuit breaker controls each branch circuit. This will allow you to turn off only power to that circuit rather than to the whole house when making repairs. Follow these procedures to identify branch circuits.

Each outlet or appliance that does not have electricity when a fuse is removed or circuit breaker is OFF is in the branch circuit controlled by that fuse or circuit breaker.

When identifying circuits, MAIN circuit breaker [3] or main switch [10] must be ON.

Removal of cartridge-type fuses [7] is explained on Page 23.

Removal of plug-type fuses and non-tamperable fuses [1] is explained on Page 23.

1. Place one circuit breaker [5] to OFF or remove one fuse [7, 1].

2. Place all wall switches [8] to ON.

3. Using a radio or small lamp, check each outlet [9] for electricity.

All major appliances such as electric ranges and washing machines are usually on separate circuits.

4. Turn all major appliances ON and OFF.

Place a label in door of box [4, 6] opposite removed fuse [7, 1] or tripped circuit breaker [5] identifying each outlet or appliance that did not have electricity.

5. Place circuit breaker [5] to ON or install fuse [7, 1]. If installing screw-in breaker [2], push in fuse button.

6. Repeat Steps 1 through 5 to identify circuits for all circuit breakers or fuses.

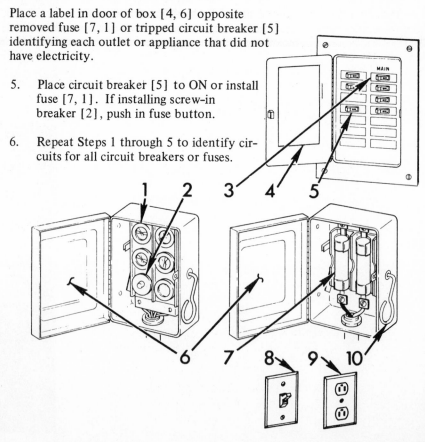

Whenever a fuse blows, your first step is to restore power as soon as possible. Your next step is to determine the problem.

Follow the procedures below if you have a defective cartridge fuse [2]. For plug-type fuses [5], go to the next section. For resetting screw-in breakers [6], go to Page 24.

▶ Replacing Cartridge Fuses

WARNING

You can be seriously injured by electricity. Do not touch any wires. If ground or floor is damp, place boards over the area and stand on boards when removing or replacing fuses.

1. Place main switch [4] to OFF. Open fuse box cover.

2. Locate blown fuse [2].

3. Place all wall switches to OFF and remove electrical plugs from all outlets in this circuit.

Refer to the label inside fuse box cover for switches and outlets in the circuit controlled by blown fuse. Page 22 describes how to identify circuits controlled by each fuse.

WARNING

Metal end caps [1] may be hot. When removing fuse let fuse fall to floor or ground.

4. Using a wooden stick, pry blown fuse from holder [3].

Number on fuse [2] indicates current rating of fuse. When replacing blown fuse, BE SURE number on new fuse is the same as number on old fuse.

5. Press new fuse into holder [3]. Close fuse box cover. Place main switch [4] to ON.

6. Go to Page 24 for circuit checkout.

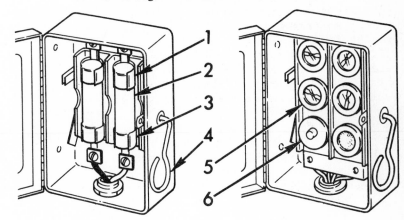

▶ Replacing Plug-Type Fuses

WARNING

You can be seriously injured by electricity. Do not touch any wires. If ground or floor is damp, place boards over the area and stand on boards when removing or replacing fuses.

1. Open fuse box cover.

If your fuses are type [1] or [6], continue.

If your fuses are reset type [3], go to Page 24.

If glass window [2] on any fuse [1] is blackened, a short circuit caused your fuse to blow.

Look for melted wire [5] or loose spring end [4] inside glass window [2]. Fuse in this condition indicates a circuit overload.

Refer to the label inside fuse box cover for switches and outlets in the circuit controlled by blown fuse. Page 22 describes how to identify circuits controlled by each fuse.

2. Place all wall switches to OFF and remove electrical plugs from all outlets in this circuit.

3. Remove blown fuse [1] by turning counter-clockwise.

Number on end of fuse [1, 3] indicates current rating of fuse. When replacing blown fuses, BE SURE number on new fuse is the same as number on old fuse.

4. Install fuse [1] in socket. Tighten fuse by turning clockwise.

5. Close fuse box cover.

6. Go to Page 24 for circuit checkout.

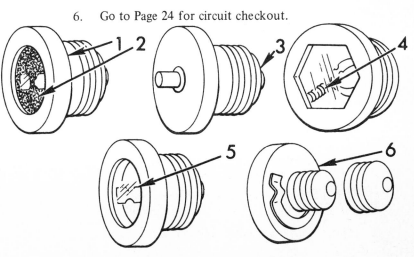

REPLACING FUSES

▶ **Resetting Screw-In Breakers**

Reset buttons [1] on fuses pop out, or trip to break circuit when overloaded.

1. Open fuse box cover.

2. Check buttons on fuses. Locate button that has tripped.

Refer to the label inside fuse box cover for switches and outlets in the circuit controlled by fuse. Page 22 describes how to identify circuits controlled by each fuse.

3. Place all wall switches to OFF and remove electrical plugs from all outlets in this circuit.

4. Push in fuse button [1]. Close fuse box cover.

5. See next section (below) for circuit checkout.

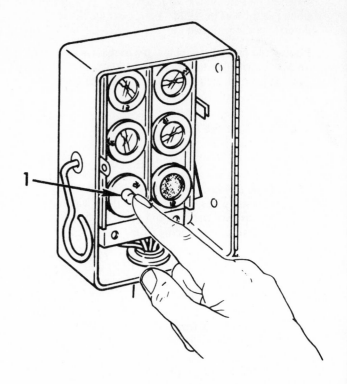

▶ **Circuit Checkout**

A circuit checkout should be performed each time a fuse is replaced. The checkout determines whether an appliance or switch in the circuit is defective, whether the circuit is overloaded, or whether the old fuse itself was defective.

Insert electrical plugs into outlets [1] one at a time. Place switches [2] to ON one at a time.

The appliance that was plugged in or the switch that was turned on when fuse [3] blows or button [4] trips is defective.

The appliance or fixture should be repaired.

If fuse [3] blows or button [4] trips only when all switches are turned on or all appliances are plugged in and turned on, one or more appliances will have to be removed from that fuse's circuit to reduce the electrical load.

If fuse [3] does not blow or button [4] does not trip a second time, you have solved the problem.

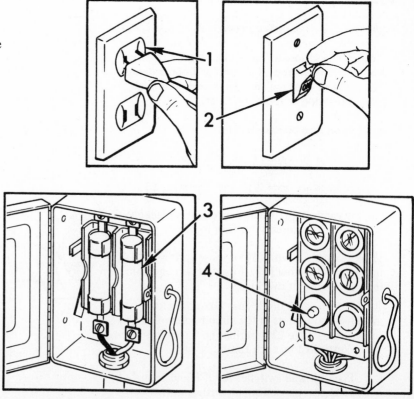

Whenever a circuit breaker trips, your first step is to restore power as soon as possible. Your next step is to determine the problem. Follow the following procedure if you have a tripped circuit breaker.

1. Open cover of circuit breaker box.

2. Check for circuit breaker [1] in middle or OFF position or for push button [2] which sticks out farther than other push buttons.

Refer to the label inside circuit breaker door to determine switches and outlets in the circuit controlled by this circuit breaker. Page 22 describes how to identify circuits controlled by each circuit breaker.

3. Place all wall switches to OFF and remove electrical plugs from all outlets controlled by circuit breaker.

4. Place circuit breaker [1] to RESET or full OFF position, then to ON position.

If circuit breaker [1] does not remain in ON position, wait one minute to repeat Step 4.

If circuit breaker remains in ON position, close circuit breaker box cover. See next section (below) for circuit checkout.

If circuit breaker [1] still does not remain in ON position, you should call an electrical repairman.

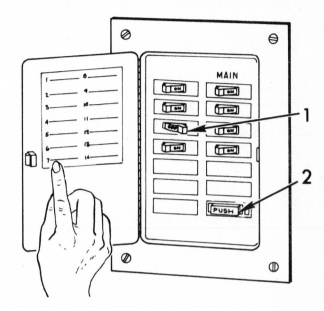

▶ **Circuit Checkout**

A circuit checkout should be performed each time a circuit breaker is reset. The checkout determines whether an appliance or switch in the circuit is defective, whether the circuit is overloaded, or whether the circuit breaker itself is defective.

Insert electrical plugs into outlets [1] one at a time. Place all switches [2] to ON one at a time.

The appliance that was plugged in or the switch that was turned on when circuit breaker [3] or push button [4] "trips" or goes to OFF is defective.

The appliance or fixture should be repaired.

If circuit breaker [3] or push button [4] "trips" or goes to OFF only when all switches are turned on or all appliances are plugged in and turned on, one or more appliances will have to be removed from that circuit breaker's circuit to reduce the electrical load.

If circuit breaker [3] remains at ON position, you have solved the problem.

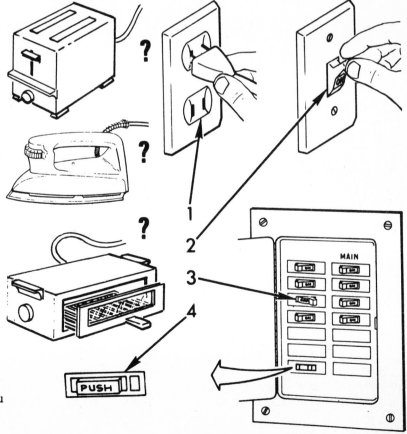

Wire connections that are physically and electrically sound are essential for safe and reliable electrical repairs and installations.

The following tools and supplies may be required to make wire connections:

Wire strippers [1] or knife [2]
Long nose pliers [3]
Diagonal cutting pliers [4]
Common screwdriver [5]
Crimping tool [6]. Obtain a quality multi-purpose crimp tool. It can be used for wire stripping as well as crimping.
Soldering gun [7] or soldering iron [8]
Resin-core solder. Be sure that solder is resin core, not acid core. Acid-core solder must not be used for electrical repairs. It can corrode copper and damage electrical parts.
Electrical tape. Vinyl plastic tape is best for home uses. It insulates well and waterproofs well. Because it is thin, it permits neat rather than bulky connections and repairs.

WARNING

Be sure that electrical power is OFF before beginning electrical repairs or connections.

The following procedures are described:

Preparing Wires for Connection, below.
Installing Crimp-Type Connectors, Page 27.
Connecting Wires to Screw Terminals, Page 27.
Connecting Wires with Solderless Connectors, Page 28.
Splicing Wires, Page 28.
Tapping Wires, Page 30.

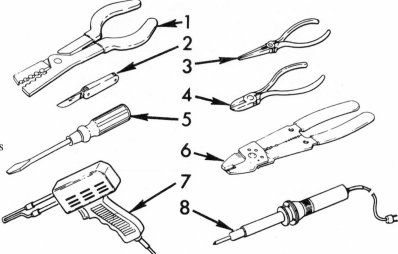

▶ **Preparing Wires for Connection**

Insulation must be removed from the wire. The length of insulation to be removed depends upon the type of connection to be made. The correct length is provided in each procedure.

If using a knife to cut insulation, cut insulation at a 60 degree angle [1] rather than straight up and down [2]. Cutting at an angle [1] reduces chances of nicking wire with knife and weakening it.

When cutting through insulation, be careful not to cut into wire.

1.   Cut through insulation around wire.

2.   Pull off insulation.

3.   Using back edge of knife, scrape wire to remove all traces of insulation.

If wire is stranded [3], be sure to clean each strand. Stranded wires should be soldered to make good connections at splices, taps and terminals. Go to Page 32 for soldering instructions.

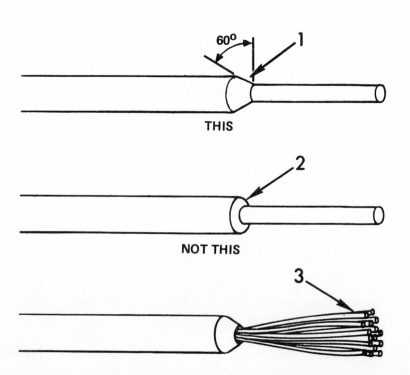

▶ Installing Crimp-Type Connectors

These procedures describe how to join (splice) wires with crimp-type splices and how to install crimp-type terminals on wires.

The following tools and supplies are required:

Crimp tool [1]. Obtain a quality multi-purpose crimp tool designed to strip wires and cut common electrical bolts as well as crimp all terminals and splices.

Crimp-type terminals [2]. Terminals must be same size (gage) as wire.

Crimp-type splices [3]. Splices must be same size (gage) as wire.

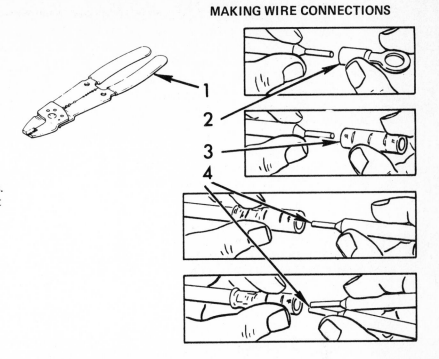

1. Remove 1/2 inch of insulation from wire, following procedures on Page 26.

2. Push terminal [2] or splice [3] onto bare wire.

Be sure that slot on crimping tool is correct size for terminal or splice.

3. Squeeze terminal or splice firmly with crimping tool until it is securely fastened to wire.

If installing splice, repeat all Steps to connect remaining wires [4].

▶ Connecting Wires to Screw Terminals

The following tools are required:

Common screwdriver [1]
Long nose pliers [2]

If wires are stranded, they should be soldered before attaching to screw terminals. Go to Page 32 for soldering instructions.

Only enough insulation should be removed to make a loop of bare wire around the terminal [3]. If more insulation is removed, bare wire [4] will be exposed. Electrical hazard will result.

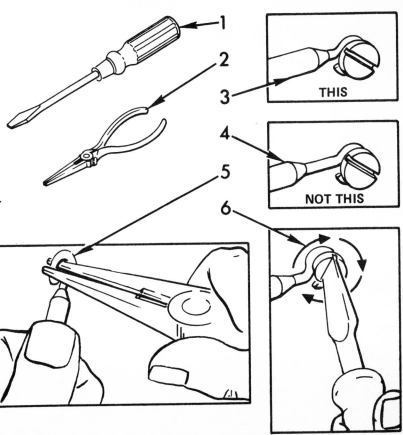

1. Remove about 1/2 inch insulation from wire, following procedures on Page 26.

2. Using pliers, bend wire to form a loop [5].

Loop [5] must be positioned around screw [6] so that tightening the screw will close the loop.

3. Place loop [5] around screw [6].

4. Using pliers, close loop [5] completely.

5. Tighten screw [6] by turning clockwise.

MAKING WIRE CONNECTIONS

▶ **Connecting Wires with Solderless Connectors**

Solderless connectors are used only if there will be no strain on the connection. They are commonly used in electrical boxes [2] where wires are protected from being pulled.

For greatest convenience, use an insulated solderless connector [1].

Size of connector must match size of wire.

If one wire is smaller in size than another wire, remove about 1/4 inch more insulation from smaller wire.

1. Remove about 3/4 inch insulation from wires, following procedures on Page 26.

2. Twist wires [3] together.

3. Install connector [4] by pushing it against end of wires and turning it clockwise.

4. Check that bare wire does not extend from end of connector. Cover with electrical tape if required.

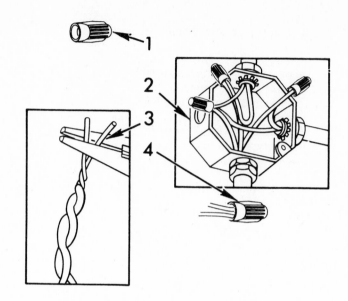

▶ **Splicing Wires**

Before splicing wires, be sure to check your local electrical codes. In some cases, connections between wires of different sizes are prohibited. In other cases, splicing may not be permitted.

There are three rules for splicing wires to obtain a sound connection:

● Connections must be as strong mechanically as a continuous length of wire.

● Connections must be as good electrically as a continuous length of wire.

● Connections must be insulated equivalent to the thickness of the original insulation.

The following tools and supplies are required:

   Wire stripper [1] or knife
   Long nose pliers [2]
   Soldering iron [3] or soldering gun
   Resin-core solder
   Vinyl plastic electrical tape

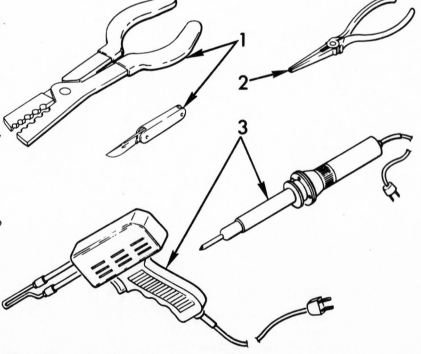

Splicing Wires

When preparing wires for splicing, it is a good idea to taper the insulation [1]. A long taper will provide a good surface for applying the electrical tape to make a smooth, even splice. Taper should be about 20 degrees.

If wire is stranded, see next section (below).

If wire is not stranded, continue.

If wire has two layers of insulation, cut outer insulation [2] back farther than inner insulation [3].

1. Remove about 3 inches of insulation from wires, following procedures on Page 26.

2. Cross wires [4] about 1 inch from insulation.

3. Twist wires [5] tightly around each other.

4. Go to Page 32, Step 3 for instructions on soldering and insulating the connection.

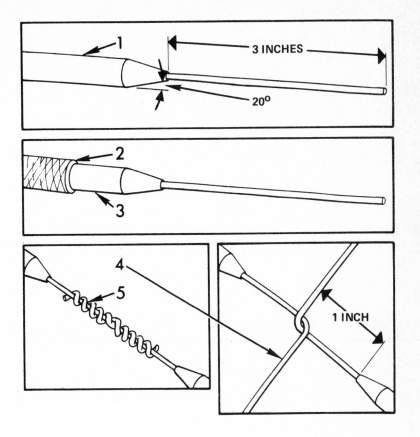

Splicing Wires

If wire has two layers of insulation, cut outer insulation [1] back farther than inner insulation [2].

1. Remove about 3 inches of insulation from wires, following procedures on Page 26.

2. Spread strands [3] evenly.

3. Place wires together so that strands [4] cross.

4. Tightly wrap one strand [5] of wire A around all strands of wire B.

5. Twisting in opposite direction, tightly wrap one strand [6] of wire B around all strands of wire A.

6. Repeat Steps 4 and 5 until all strands [7] are wrapped.

7. Go to Page 32, Step 3 for instructions on soldering and insulating the connection.

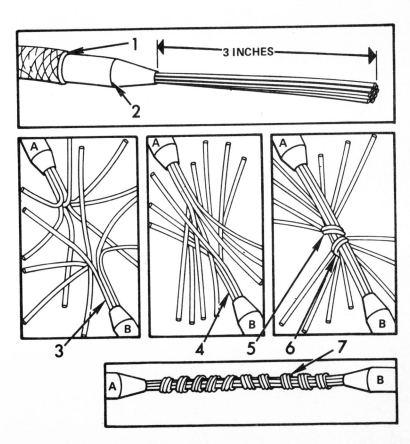

MAKING WIRE CONNECTIONS

▶ **Tapping Wires**

The procedure for connecting a wire [5] to a continuous length of wire is called tapping.

Tapping must not be done if there will be any strain or pull on the connection.

To avoid the necessity of stripping wires, consider using self-stripping electrical tap connectors. They are available in several sizes to permit tapping different sized wires. They are convenient to use and are especially suitable for use in mobile homes, trailers and other low-voltage systems. Only a pliers is required to join or tap the wires.

Check local codes for application to home wiring.

When stripping wires for tapping, it is a good idea to taper [4] the insulation. A long taper will provide a good surface for applying the electrical tape to make a smooth, even connection. Taper should be about 20 degrees.

The following tools and supplies are required:

    Wire stripper [1] or knife
    Long nose pliers [2]

Soldering iron [3] or soldering gun
Resin-core solder
Vinyl plastic electrical tape

If wire is stranded, go to Page 31.
If wire is not stranded, go to next section (below).

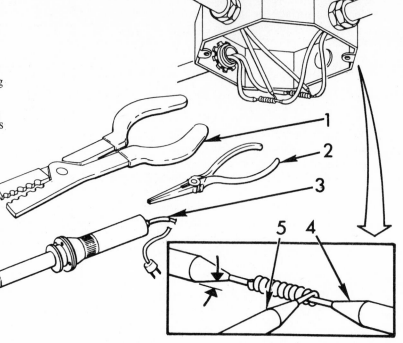

Tapping Wires

If wire has two layers of insulation, cut outer insulation [1] back farther than inner insulation [2].

1. Remove about 1 inch of insulation from wire [3] to be tapped. Follow procedures on Page 26.

2. Remove about 1-1/2 inches of insulation from other wire [4].

3. Beginning 1/4 inch from insulation, twist wire [5] tightly around tapped wire.

4. Go to Page 32, Step 3 for instructions on soldering and insulating the connection. Be sure to insulate all exposed wire.

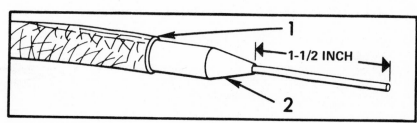

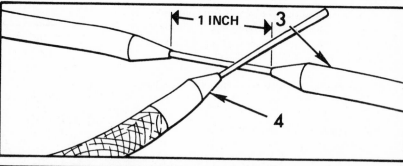

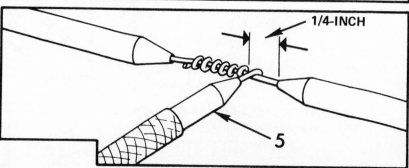

Tapping Wires

If wire has two layers of insulation, cut outer insulation [1] back farther than inner insulation [2].

1. Remove about 2 inches of insulation from wire [3] to be tapped. Follow procedures on Page 26.

2. Remove about 3 inches of insulation from other wire [4].

3. Separate strands [5] of wire into two halves.

4. Insert wire [6] between halves. Separate strands of wire [6] into two equal groups.

5. Tightly wrap one group of strands [7] around wire [8] in clockwise direction.

6. Tightly wrap other group of strands around wire in counterclockwise direction.

7. Go to Page 32, Step 3 for instructions on soldering and insulating the connection. Be sure to insulate all exposed wire.

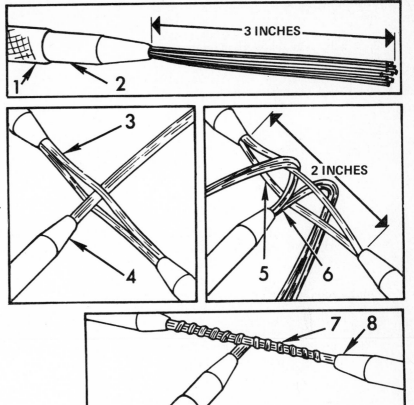

HOW TO SOLDER

Wire connections are frequently soldered to improve their physical strength and electrical contact.

Solder is most commonly used to secure wire and electrical parts to terminals. It is also used to secure wire splice and tap connections.

Good soldering is not difficult if a few principles are followed. Practice is required to develop the technique of flowing solder into the electrical connections.

The following tools and supplies are required:

>   Knife [1]
>   Diagonal cutting pliers [2]
>   Soldering iron [5] or soldering gun [4]
>   Fine-tooth file [6] or sandpaper. This equipment is used to clean the tip of the soldering copper [3].
>   Resin-core solder. Be sure that solder is resin core, not acid core. Acid-core solder must not be used for electrical repairs. It can corrode copper and damage electrical parts.
>   Electrical tape. Plastic tape is best for home use. It insulates well and waterproofs well. Because it is thin, it permits neat rather than bulky connections.

WARNING

Be sure that electrical power is OFF before beginning electrical repairs or connections.

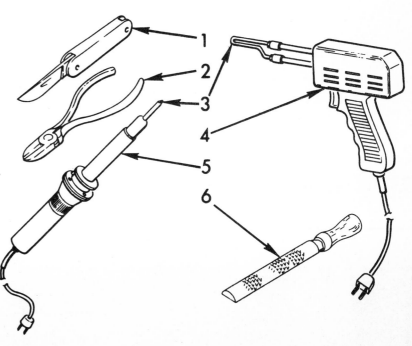

HOW TO SOLDER

Read through entire procedure before beginning task.

Be sure that wire splice or tap connection is physically strong. Page 28 and 30.

Be sure that all insulation is removed from wires and that wires are clean and shiny. Page 26.

Be sure that tip [1] of soldering copper is clean and shiny.

1. Heat soldering copper. Copper is hot enough when it easily melts solder.

2. Apply small amount of solder to tip [1]. Solder must just coat tip. This solder is used to help conduct heat from soldering copper to wire.

3. Hold tip [3] against wire [2]. Tip must be held against wire until wire becomes hot enough to melt solder. However, heat must not spread through wire and melt insulation from wire. If you have consistent trouble with melting the insulation before the wire can melt the solder, you may need a larger soldering iron.

4. While holding tip [3] against wire [2], apply solder to wire. Begin applying the solder at a point near the tip [3]. Then while solder melts, move solder along wire. If solder stops melting, you may be moving too fast. Solder should just cover wire uniformly. If solder hangs from wire in drops, you may be using too much solder.

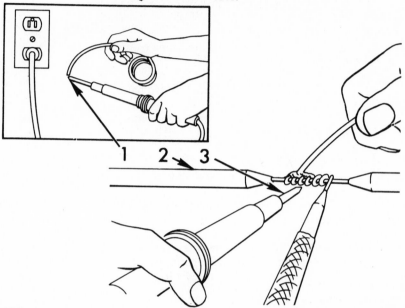

5. Remove solder.

6. Hold tip against wire until solder has flowed into spaces between wires.

7. Remove tip from wire.

8. Allow wire to cool and solder to harden. Wire must not move while solder is hardening. If wire is moved, connection may not be electrically sound or physically strong.

When applying electrical tape, be sure to keep tape stretched tightly.

9. Starting at one end [1] of connection, wind tape around wire until reaching opposite end [2] of connection. Each turn must overlap preceding turn.

10. Wind tape around wire until reaching starting point. Continue wrapping back and forth along connection until thickness of tape equals thickness of insulation.

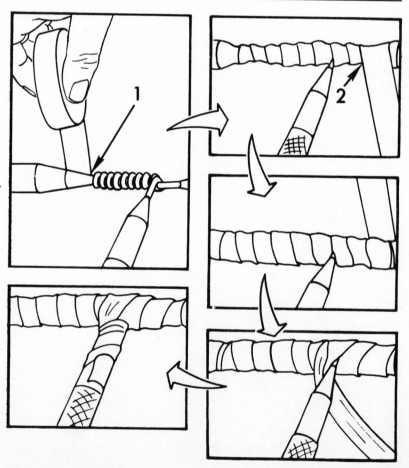

Most household extension cords and small appliance cords have only two wires. Three-wire extension cords or appliance cords are used mostly for power tools and outdoor uses. The third wire is a grounding wire.

Damaged cords are a fire hazard. Replace or repair them at once to prevent shorted circuits. If insulation on the cord is badly frayed, throw the cord away.

Sometimes extension cords are damaged by plugging in too many appliances. The wires get hot and the insulation melts. In this case, throw the cord away — it can't be repaired.

If the prongs of a plug are bent, straighten them out with pliers. If plug does not fit snuggly into outlet, carefully bend prongs outward a little. The plug will now fit correctly.

To properly care for extension cords and appliance cords, follow these tips:

- Keep cords in a cool, dry place.
- Never bend cords sharply.
- Grasp the plug when disconnecting cord — NEVER pull on the cord itself.

- When disconnecting appliance cords, disconnect at the outlet, never at the appliance.
- Keep extension cords away from door jambs and from under rugs.

Instructions on Pages 33-37 show how to replace 2- and 3-prong plugs [1, 2] and appliance cord plugs [3]. With this information you can also repair damaged cords. Simply cut off the damaged end and replace the plug.

For replacing 2- and 3-prong plugs [1, 2], see next section (below).

For replacing appliance cord plugs [3], go to Page 35.

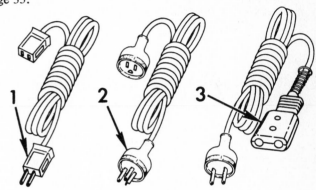

▶ Replacing 2- and 3-Prong Plugs

The following tools and supplies are required:

> Knife [1]
> Screwdriver, common or Phillips [2]
> Long nose pliers [3]
> Diagonal cutting pliers [4]

If wires are stranded:

> Resin-core solder
> Soldering gun [5] or iron [6]

Plugs are made of several different materials. Most replaceable plugs, however, are replaced in a similar manner.

Several new types of plugs [7] have been made to save replacement time. With most of these types, the cord is simply inserted into a hole in the plug, where it is clamped into place. Pins pierce the wires to make a good connection.

This type of plug [7] is time saving; however, one or two accidental pulls on the cord to unplug it will pull the wires out. For safer wiring, the older style plugs are better.

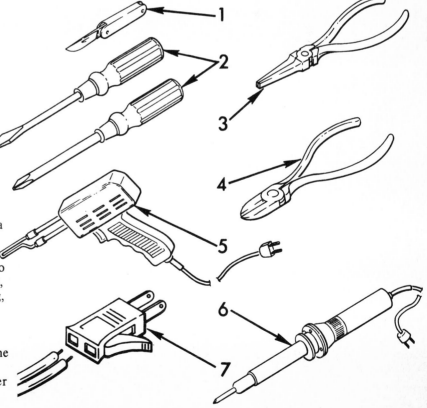

REPAIRING CORDS AND PLUGS

Replacing 2- and 3-Prong Plugs

1. Remove insulator [6] from old plug [1].

2. Loosen screws [2]. Remove wires from screws. Cut or untie knot [3], if installed. Loosen clamp [4], if installed.

3. Pull plug from cord.

If cord is not damaged, go to Step 8 to install new plug.

If cord is damaged or knot [3] was cut, continue.

4. Cut off damaged end of cord.

If cord has outer insulation, remove about 2-1/2 inches of outer insulation.

If cord is molded type [5], separate wires for a distance of about 2-1/2 inches.

5. Remove about 3/4 inch of insulation from wires. Follow procedures on Page 26.

Replacing 2- and 3-Prong Plugs

If wires are not stranded, go to Step 8 to install new plug.

If wires are stranded, continue.

6. Twist stranded wire [1] tightly.

7. Apply small amount of solder [2] to the wire. This will make it easier to attach wire to screw and keep the individual strands from shorting to other wire.

8. Insert cord through plug.

If replacing 2-prong plug, go to Page 35.

If replacing 3-prong plug, continue.

9. Tie all three wires together into a tight knot [3]. Pull cord until knot is snug against plug. Go to Page 35, Step 11.

SOLDER

Replacing 2- and 3-Prong Plugs

The best way to tie two wires together is to use the Underwriter's knot. See illustration.

10.   Tie knot [1] in wires. Pull cord until knot is snug against plug.

11.   Make loop [4] in end of each wire.

If replacing 3-prong plug, attach green wire to green or dark-colored screw.

Wires should be routed around prongs [2] before being attached to screws [3].

Be sure to position loop [4] under screw so that tightening screw will close loop.

12.   Place loop [4] under screw.

13.   Using long nose pliers, close loop completely around screw.

14.   Tighten screw.

15.   Repeat Steps 12, 13 and 14 to connect remaining wires.

16.   Install insulator [5]. Tighten clamp [6], if installed.

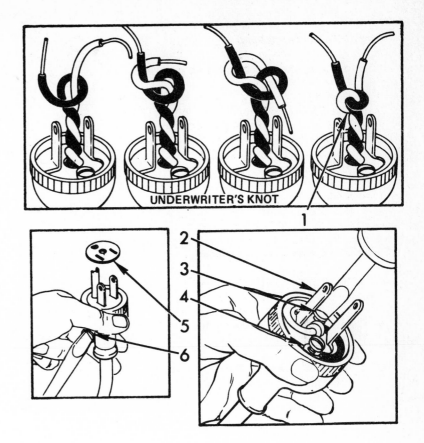

▶ Replacing Appliance Plugs and Cords

Many appliances have separate cords. These cords have a wall plug [1] on one end and an appliance plug [2] on the other end. Go to Page 33 for procedures to replace the wall plug.

The appliance plug is usually the first part to wear out. If the cord is in otherwise good shape, the appliance plug can easily be replaced. If the cord itself is worn, however, it is much safer to buy a new appliance cord rather than repair the old one.

When purchased new, most appliance cords have appliance plugs which are riveted together. They are removed by cutting the cord. Any frayed or worn cord should be removed too.

If your appliance cord has a removable appliance plug, it is a good idea to cut back the cord so that you will be working with "new" wire endings. Sometimes the appliance plug is in good condition and can be re-installed on "new" wire endings.

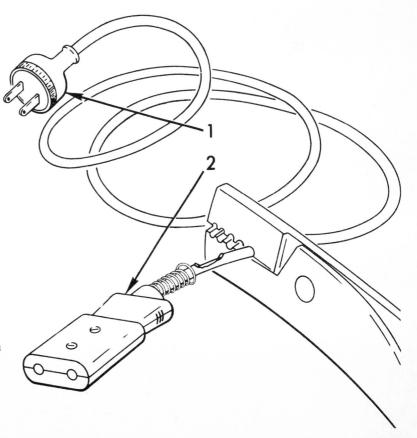

REPAIRING CORDS AND PLUGS

Replacing Appliance Plugs and Cords

The following tools and supplies are required:

> Knife [1]
> Screwdriver, common [2] or Phillips [3]
> Diagonal cutting pliers [4]
> Long nose pliers [5]

If wires are stranded:

> Soldering iron [6] or soldering gun
> Resin-core solder

If appliance plug is riveted, remove by cutting cord. Go to Step 3 to install new appliance plug.

If appliance plug is held together with screws, continue.

1. Remove casing [7] by removing screws [10] and nuts [9].

2. Loosen screws [8]. Pull cord through spring guard.

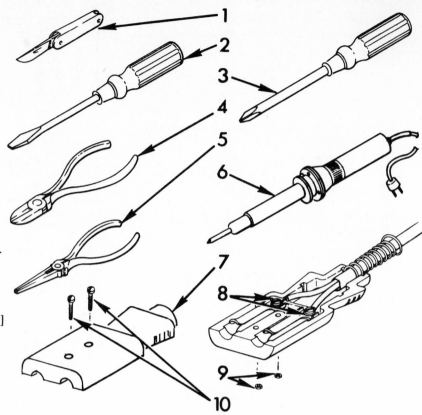

Replacing Appliance Plugs and Cords

3. Cut off end of cord. Using a knife, carefully remove about 2-1/2 inches of outer insulation.

4. Remove 3/4 inch of insulation from each wire. Follow procedures on Page 26.

If wires are not stranded, go to Step 7.

If wires are stranded, continue.

5. Twist each stranded wire [1] tightly.

6. Apply small amount of solder [2] to each wire. This will make it easier to attach wire to screw and keep the individual strands from shorting to other wires.

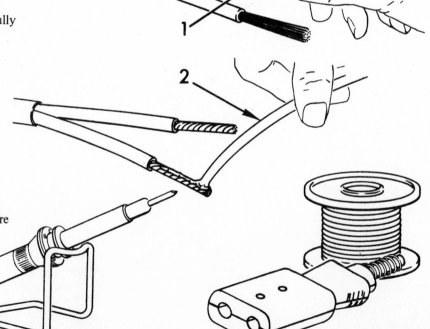

Replacing Appliance Plugs and Cords

7. Insert cord through spring guard [1].
    Bend each wire end into loop [2].

8. Place loop [2] under terminal screws [3] so
    that tightening screw (clockwise) will close
    the loop.

9. Using long nose pliers, close wire ends com-
    pletely under screws. Tighten screws.

10. Place spring guard [1] into one half of plug
    casing. Install other half of casing. Install
    screws [5] and nuts [4].

11. Plug in cord. Check operation of appliance.

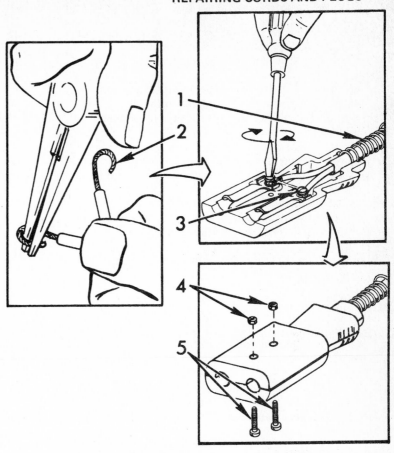

REPLACING SWITCHES

Instructions in this section show how to replace
the following types of switches:

- Common switch [1], Page 38.
- Mercury (silent) switch [1], Page 38.
- Tap switch [2], Page 38.
- Push-button switch [3], Page 38.
- Dimmer switch [4], Page 39.
- Backwired switch [5], Page 40.
- Combination switch [6], Page 41.
- 3-way and 4-way switches, Page 43.
  (Fixture controlled from two or more
  locations)
- In-line switch [7], Page 43.
- Outdoor switch [8], Page 44.

Before replacing any type of switch, read the
information on Page 38.

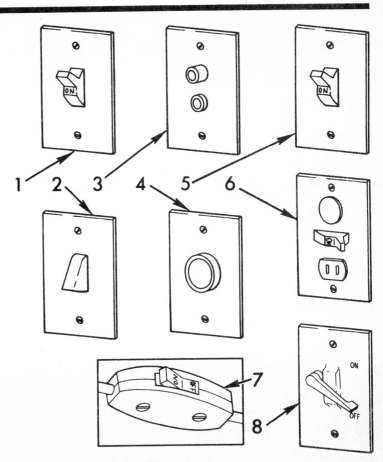

REPLACING SWITCHES

If a fuse blows or a circuit breaker trips whenever you turn on a switch, the switch is faulty and should be replaced as soon as possible.

There are many different types of switches in use today. Most switches are replaced in the same way. The difference between them is the number and location of wires.

Before replacing switches, you should look into the different types that are available. For example, you may want to replace an existing common switch with a dimmer switch. This type not only creates different lighting moods, but also will save power when it is in any position but full ON.

Cover plates are also available in several different shapes and decorations.

Switches are not difficult to replace. Just remember to put all wires back as they were. If space permits, the best way is to transfer one wire at a time from the old switch to the new switch.

Most switches made today have screws that are color-coded to the wires. Follow the chart at the right to connect wires.

Color of Wire	Color of Screw
Green	Green (or darkest screw)
Red	Brass
White	Silver
Black	Brass
Bare wire	Electrical box ground

The following tools are required to replace switches:

Screwdriver, common [1] or Phillips [2]
Long nose pliers [3]
Knife [4]

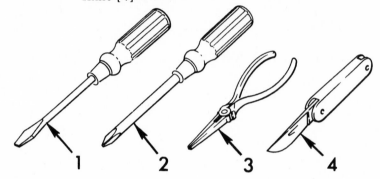

▶ Replacing Common Switches

This section describes the procedures for replacing the following types of switches:

Common switch [1]. Includes mercury (silent) switches.
Tap switch [2]
Push-button switch [3]

WARNING

Be sure to turn off circuit breaker or remove fuse that controls circuit being worked on.

If removing a cover plate [5] from a wall that is painted, carefully cut around edge of plate with a razor blade or a sharp knife to prevent paint from peeling off wall.

1. Remove screws [4]. Remove cover plate [5].

2. Remove mounting screws [6]. Pull switch [7] from box.

3. Loosen screws [9]. Remove wires [8] from screws. Remove switch [7].

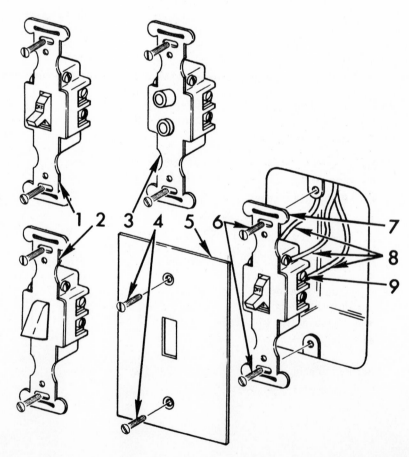

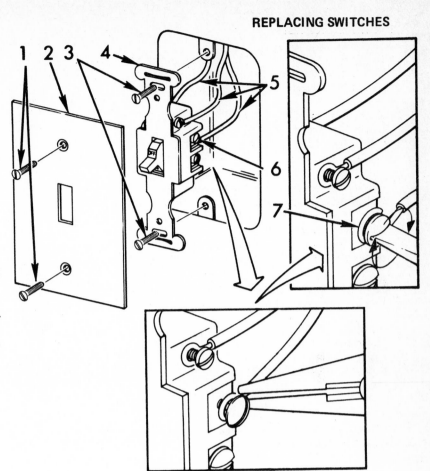

Replacing Common Switches

4. Install wires [5] on screws [6] of new switch [4]. Position wires around screws so that as screw is tightened, the loop [7] in end of wire is closed.

5. Using long nose pliers, tighten loop [7] around screws [6]. Tighten screws.

6. Place switch [4] in box. Install mounting screws [3].

7. Install cover plate [2]. Install screws [1].

8. Place circuit breaker to ON or install fuse.

9. Check operation of switch [4].

► Replacing Dimmer Switch

Dimmer switches [1] can replace almost any ordinary switch. By using a dimmer switch, you can cut down on electricity consumption, thereby reducing your utility bill.

When buying a dimmer switch, be sure its wattage rating is higher than the total wattage of the lights being controlled by the switch.

WARNING

Be sure to turn off circuit breaker or remove fuse that controls circuit being worked on.

If removing cover plate [3] from a wall that is painted, carefully cut around edge of plate with a razor blade or sharp knife to prevent paint from peeling off wall.

If replacing a common switch [2] with a dimmer switch, go to Page 38 to remove common switch. After removal, go to Page 40, Step 5 to install dimmer switch.

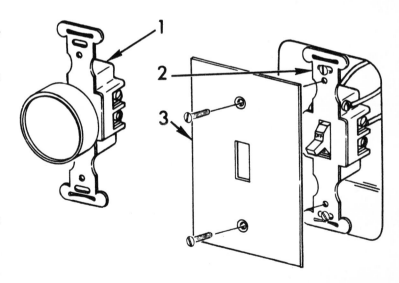

REPLACING SWITCHES

Replacing Dimmer Switch

1. Loosen set screw [9], if installed. Pull off dimmer knob [10].

2. Remove screws [1]. Remove cover plate [2].

3. Remove mounting screws [3]. Pull switch [4] from box [5].

4. Loosen screws [7]. Remove wires [6]. Remove switch [4].

5. Install wires [6] on screws [7] of new switch [4]. Position wires around screws so that as screw is tightened, the loop [8] in end of wire is closed.

6. Using long nose pliers, tighten loops [8] around screws [7]. Tighten screws.

7. Place switch [4] in box [5]. Install mounting screws [3].

8. Install cover plate [2]. Install screws [1].

9. Press on knob [10]. Tighten set screw [9] if required. Turn knob to OFF position.

10. Place circuit breaker to ON or install fuse.

11. Check operation of dimmer switch.

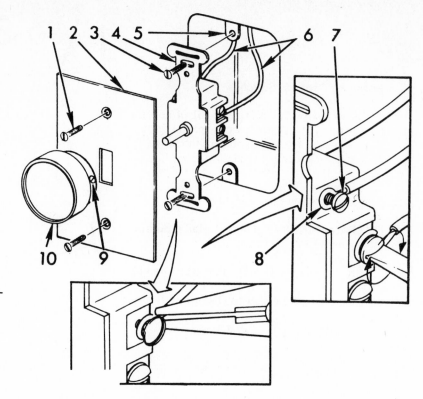

▶ **Replacing Backwired Switch**

Backwired switches do not require screws for connecting wires. Wires are held automatically when inserted into slots [6] in the back of the switch. Backwired switches can replace most other types of common switches.

WARNING

Be sure to turn off circuit breaker or remove fuse that controls circuit being worked on.

If removing cover plate [2] from a wall that is painted, carefully cut around edge of plate with razor blade or sharp knife to prevent paint from peeling off wall.

1. Remove screws [1]. Remove cover plate [2].

2. Remove mounting screws [3]. Pull switch [4] from box [5].

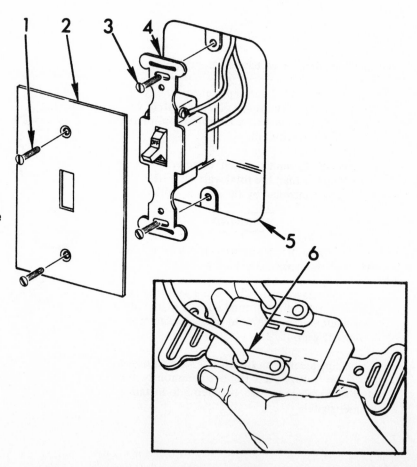

Replacing Backwired Switch

If replacing existing backwired switch [4], go to Step 4.

If removing common switch [6] without backwiring, continue.

3. Loosen screws [7]. Remove wires [8] from switch [6]. Remove switch. Go to Step 5.

4. Locate wire release [11]. While pulling wire [10] from switch, press release with screwdriver.

5. Straighten and clean ends of wires. Using strip guide [13] on switch casing, trim off insulation or cut wires as required.

6. Insert wires [9] in slots [12]. Observe color guide on back of casing if connecting more than two wires.

7. Place switch in box. Install mounting screws [3].

8. Install cover plate [2]. Install screws [1].

9. Place circuit breaker to ON or install fuse. Check operation of switch.

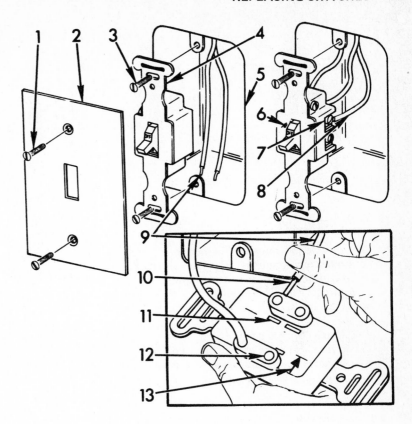

▶ **Replacing Combination Switches**

Combination switches [1] can consist of numerous combinations of switches, lights or outlets. However, the instructions for replacing most of these types are the same.

Combination switches [1] may or may not require a jumper wire [3] between different parts of the switch. If wire is required, be sure it is the same gage as the circuit wiring [4].

If replacing a common switch [2] with a combination switch [1], go to Page 38 to remove common switch.

WARNING

Be sure to turn off circuit breaker or remove fuse that controls circuit being worked on.

If removing cover plate from a wall that is painted, carefully cut around edge of plate with razor blade or sharp knife to prevent paint from peeling off wall.

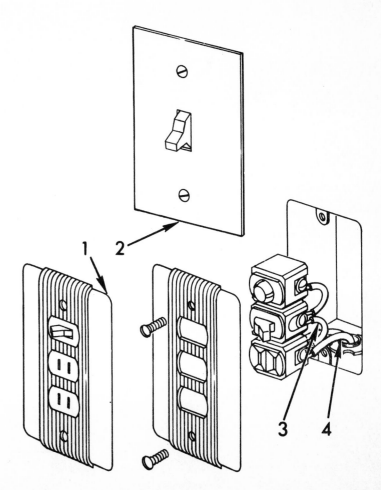

REPLACING SWITCHES

Replacing Combination Switches

1. Remove screws [1]. Remove cover plate [2].

2. Remove mounting screws [3]. Pull switch [4] from box [5].

3. Loosen screws [8]. Remove wires [7].

If jumper wire [6] is not required for new switch, go to Step 5. If wire is required, continue.

Jumper wire [6] must be trimmed to connect between screw terminals [8]. Trim about 1/2 inch of insulation from each end and middle of wire if required. Page 27 describes connecting wires to screw terminals.

4. Connect jumper wire [6] to screw terminals [8] of new switch. Position ends of wires around screws so that as screws are tightened, loops are closed.

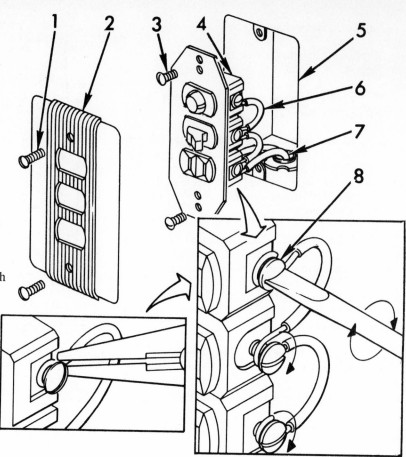

Replacing Combination Switches

5. Install wires [6] on screws [8] of new switch [4]. Position wires around screws so that as screw is tightened, loops [7] at end of wires are closed.

6. Using long nose pliers, close loops [7] around screws [8]. Tighten screws.

7. Place switch [4] in box [5]. Install mounting screws [3].

8. Install cover plate [2]. Install screws [1].

9. Turn circuit breaker to ON or install fuse.

10. Check operation of switch.

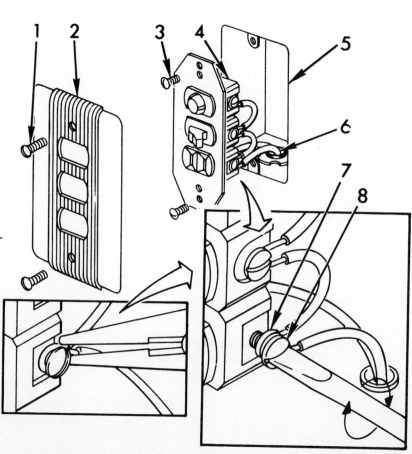

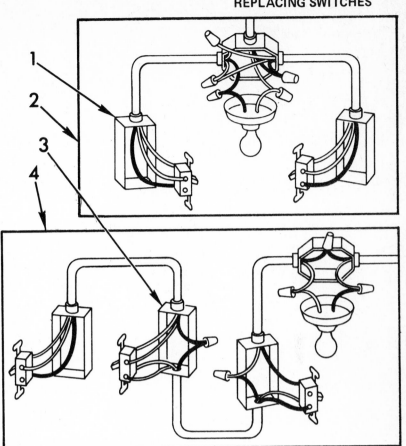

▶ Replacing 3-Way and 4-Way Switches

A 3-way switch [1] is one that controls a fixture from two separate locations [2] such as the top and bottom of a stairway.

A 4-way switch [3] is needed whenever you wish to control a switch from three separate locations [4]. The 4-way switch is placed between two 3-way switches for this purpose.

A 3-way or 4-way switch is replaced in the same way as a common switch except that more wires are connected. Use the illustration of 3-way switch [1] or 4-way switch [3] as a guide for position of wires.

Follow instructions on Page 38 for replacement. Label removed wires for aid during installation.

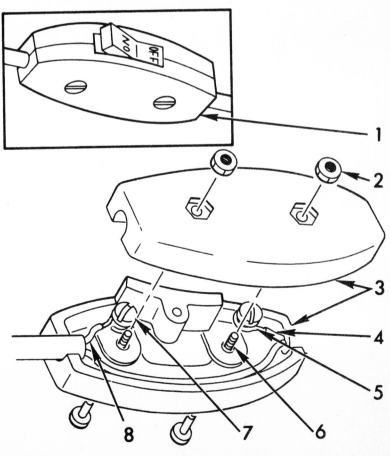

▶ Replacing In-line Switch

An in-line switch [1] is basically a controllable splice in a cord.

WARNING

Be sure to disconnect cord being worked on from power source.

1. Remove two screws [6] and nuts [2]. Separate switch halves [3].

If installing new switch on continuous length of wire, go to Page 44, Step 3.

If removing old switch, continue.

2. Loosen screws [5, 7]. Disconnect wires [4, 8]. Remove old switch. Go to Page 44, Step 5.

REPLACING SWITCHES

Replacing In-line Switch

3. Separate the two wires [3, 2] in cord [1]. Cut one wire [3] only.

4. Trim off about 1/2 inch of insulation from cut ends of wire [3]. Follow the procedures on Page 26.

5. Bend ends of wire [3] into loops [4].

6. Place continuous wire [2] into slot [8] in switch.

7. Install wires [3] on screws [7, 10] of new switch. Position wires [3] around screws so that as screw is tightened, the loop [4] in end of wires is closed.

8. Tighten loops [4] around screws [7, 10] with long nose pliers.

9. Tighten screws [7, 10].

10. Place switch halves [6] together. Install two screws [9] and nuts [5].

11. Plug in cord. Check operation of switch.

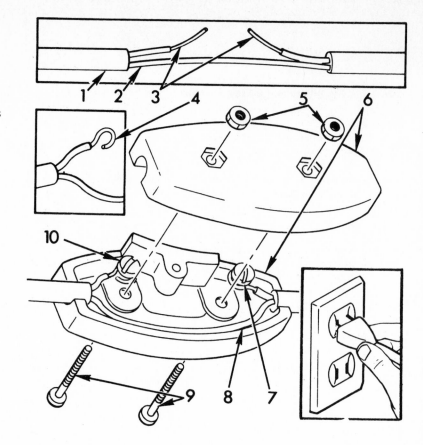

▶ Replacing Outdoor Switch

The basic difference between indoor and outdoor switches is the cover plate [2]. Outdoor switches must be protected from the weather by a gasket [1] and weatherproof cover plate.

WARNING

Be sure to turn off circuit breaker or remove fuse that controls circuit being worked on.

1. Remove screws [11]. Remove cover plate [10] and gasket [9].

2. Remove mounting screws [8]. Pull switch [6] from box [3].

3. Loosen screws [5]. Remove wires [4]. Remove old switch [6].

4. Install wires [4] on screws [5] of new switch [6]. Position wires around screws so that as screws are tightened, the loops [7] in ends of wires are closed.

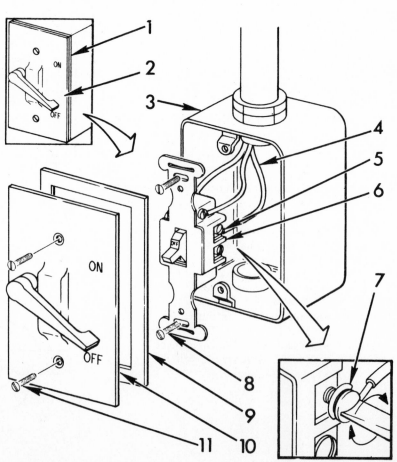

Replacing Outdoor Switch

5. Tighten loop [1] around screws [2] with long nose pliers.

6. Tighten screws [2].

7. Place switch [3] in box [4]. Install mounting screws [6].

8. Check gasket [8] for damage. Replace as required.

9. Place switch [5] to OFF. Place lever [7] on cover plate [9] to OFF.

10. Install gasket [8]. Install cover plate [9]. Install screws [10].

11. Turn circuit breaker to ON or install fuse. Check operation of switch.

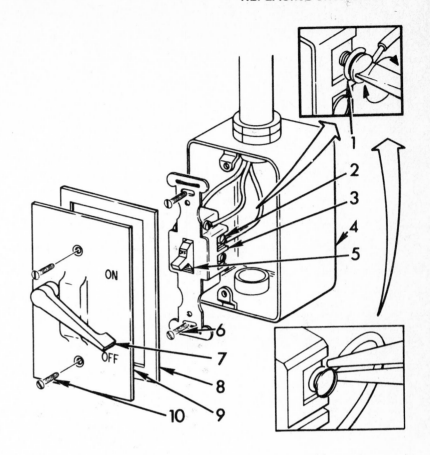

REPLACING OUTLETS

If an electrical outlet, or receptacle, is faulty, it will short the circuit whenever any appliance or lamp is plugged into it. It should be replaced as soon as possible.

There are many different types of outlets in use today. Most are replaced in the same manner. The difference between them is the location and number of wire connections.

The Code was recently changed so that in new construction, or whenever replacing an old outlet, it must be replaced with the grounding type outlet [7], both indoors and outdoors. This type will accommodate both 2-prong plugs [5] and 3-prong plugs [6]. Check with your local codes before changing or installing outlets.

The instructions in this section show how to replace the following types of outlets:

Convenience outlets, Page 46.

Backwired outlets, Page 47.

Outdoor outlets, Page 48.

The following tools are required:

Knife [1]
Long nose pliers [2]
Screwdriver, common [3] or Phillips [4]

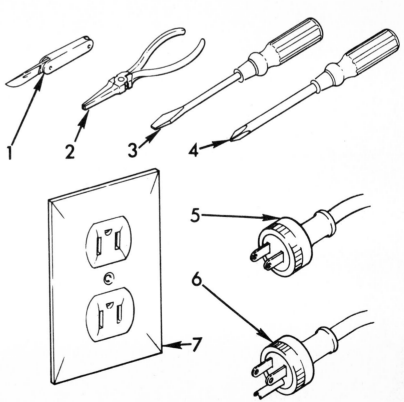

45

REPLACING OUTLETS

▶ Replacing Convenience Outlets

Single receptacle outlets [1] are normally not in use today. If you are replacing one, you should consider a double receptacle [2].

There are many different combinations of switches, lights, and receptacles that you may want to install.

For example, if you have small children, you may want to install a safety outlet [3]. A safety outlet has plastic disks which must be turned before a plug can be inserted into the receptacle. This requires more strength and skill than a child normally has.

Before replacing an outlet, look into other styles. You may want to change styles.

Whenever you are replacing an outlet with screw connections, the best procedure is to transfer one wire at a time from the old outlet to the new outlet, if space permits.

Most outlets made today have screws that are color-coded to the wires. Follow the chart at the right to connect wires.

Color of Wire	Color of Screw
Green	Green (or darkest screw)
Red	Brass
White	Silver
Black	Brass
Bare wire	Electrical box ground

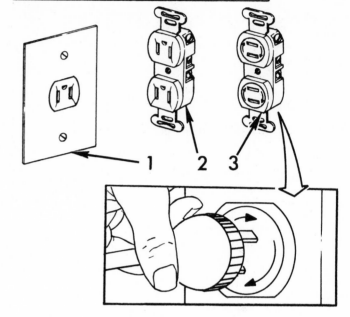

Replacing Convenience Outlets

WARNING

Be sure to turn off circuit breaker or remove fuse that controls circuit being worked on.

If removing a cover plate [2] from a wall that is painted, carefully cut around edge of plate with razor blade or sharp knife to prevent paint from peeling off wall.

1. Remove screw(s) [1]. Remove cover plate [2].

2. Remove mounting screws [3]. Pull outlet [4] from box [5].

3. Loosen screws [7]. Remove wires [6] from screws. Remove outlet [4].

4. Install wires [6] on screws [7] of new outlet [4]. Position wires around screws so that as screws are tightened, the loops [8] in ends of wires are closed.

5. Tighten loop [8] around screws [7] with long nose pliers.

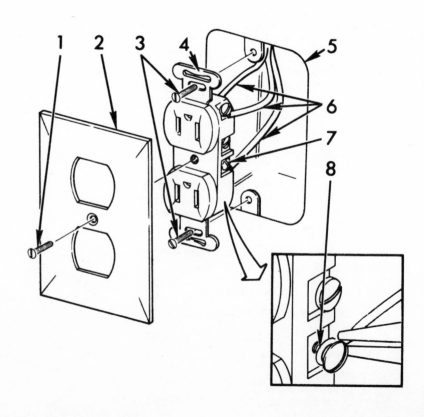

Replacing Convenience Outlets

6. Tighten screws [6].

7. Place outlet [4] in box [5]. Install mounting screws [3].

8. Install cover plate [2]. Install screw(s) [1].

9. Place circuit breaker to ON or install fuse.

10. Insert electrical plug of appliance or lamp into receptacle. Check operation of receptacle.

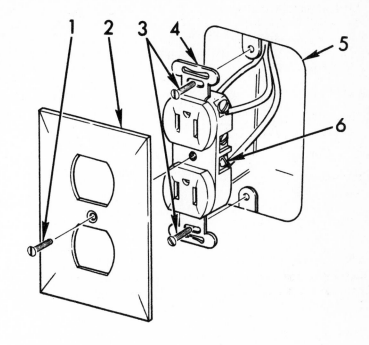

▶ Replacing Backwired Outlets

Backwired outlets do not require screws for connecting wires. Wires are held automatically when inserted into slots [9] in the back of the outlet. Backwired outlets can replace most outlets with screw terminals.

WARNING

Be sure to turn off circuit breaker or remove fuse that controls circuit being worked on.

If removing a cover plate [2] from a wall that is painted, carefully cut around edge of plate with razor blade or sharp knife to prevent paint from peeling off wall.

1. Remove screw [1]. Remove cover plate [2].

2. Remove mounting screws [3]. Pull outlet [4 or 8] from box [5].

If replacing existing backwired outlet [8], go to Page 48, Step 4.

If replacing outlet [4] without backwiring, continue.

3. Loosen screws [7]. Remove wires [6] from screws. Remove outlet [4]. Go to Page 48, Step 5.

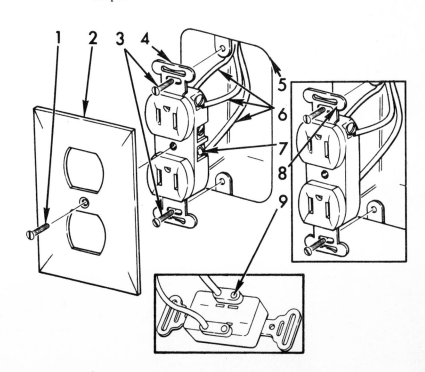

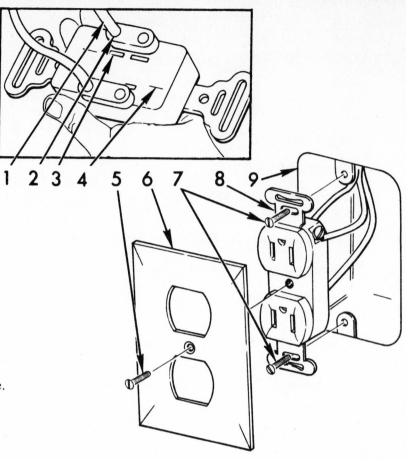

Replacing Backwired Outlets

4. Locate wire release [3]. While pulling wires [1] from outlet, press release with screwdriver.

5. Straighten and clean wires [1]. Using strip guide [4] on outlet casing, trim off insulation or cut wires as required.

6. Insert wires in slot [2]. Observe color guide on back of outlet if connecting more than two wires.

7. Press outlet [8] in box [9]. Install mounting screws [7].

8. Install cover plate [6]. Install screw [5].

9. Place circuit breaker to ON or install fuse.

10. Insert electrical plug of appliance or lamp into receptacle. Check operation of receptacle.

▶ **Replacing Outdoor Outlets**

The basic difference between indoor and outdoor outlets is the cover plate [2]. Outdoor outlets must be protected from the weather by a gasket [1] and weatherproof cover plate.

All outdoor outlets should be protected by a ground-fault circuit interrupter (GFCI) [3]. This device detects very small amounts of current-to-ground leakage which regular circuit breakers or fuses cannot detect. This eliminates the undetected presence of a dangerous shock.

Some GFCI's are installed just as you would install an outlet; other models are installed in several different ways. They can be wired to protect an entire circuit of outlets or just one outlet. Be sure to follow manufacturer's instructions when installing a GFCI.

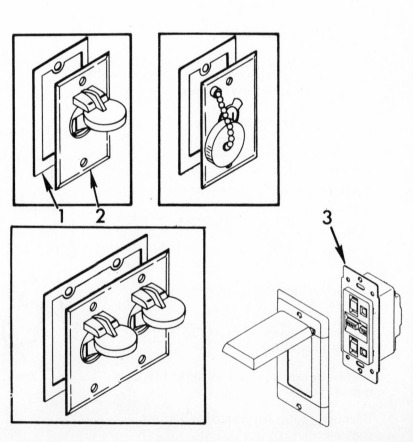

Replacing Outdoor Outlets

The following tools will be required to replace an outdoor outlet with a new outlet or a GFIC [8]:

Screwdriver, common [1] or Phillips [2]
Long nose pliers [3]

WARNING

Be sure to turn off circuit breaker or remove fuse that controls circuit being worked on.

1.   Remove screws [4]. Remove cover plate [5] and gasket [6].

2.   Remove mounting screws [7]. Pull outlet from box [9].

3.   Remove wires from outlet. Remove outlet.

4.   Attach wires to new outlet or GFCI [8]. If installing new outlet, go to Step 5. If installing GFCI, follow manufacturer's installation instructions. Then go to Step 8.

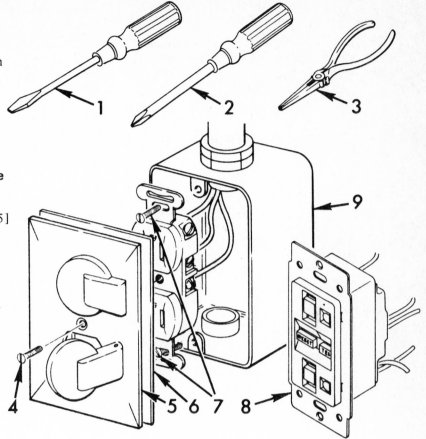

Replacing Outdoor Outlets

5.   Place outlet [5] in box [6]. Install mounting screws [4].

6.   Check gasket [2] for damage. Replace if required.

7.   Install gasket [2]. Install cover plate [3]. Install screws [1].

8.   Place circuit breaker to ON or install fuse.

9.   Insert electrical plug of appliance or lamp into receptacle. Check operation of receptacle.

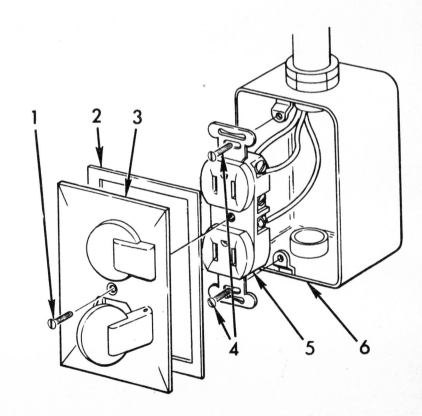

Most homes have a number of lighting fixtures, both wall and ceiling, along with floor and table lamps.

For repair of ceiling and wall fixtures, both incandescent and fluorescent, go to Page 52.

For repair of portable lamps, continue.

▶ Repairing Lamps

The main parts of a lamp that wear out are:

    Bulb [1]
    Socket [2] and switch [3]
    Cord and plug [4]

If the cord or socket do wear and are not noticed, the lamp becomes a dangerous fire or shock hazard.

If the lamp will not go on or if it flickers on and off, check that the circuit is getting power and the plug is properly plugged in. Also be sure the bulb is good.

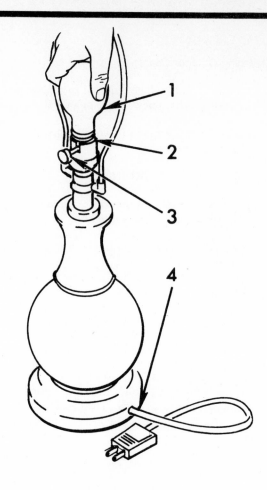

Repairing Lamps

If the bulb is not good, replace it. If you cannot remove the bulb because the glass is broken, simply wad up a sheet of newspaper or paper sack tightly. Then firmly press down on broken bulb and turn counterclockwise until base of bulb is removed. Discard the paper and base.

If replacing the bulb does not solve the problem, you will need to repair the lamp. There are many different types of lamps. They may differ in construction but they are electrically the same.

The following tools and supplies are required:

    Screwdriver, common [1] or Phillips [2]
    Long nose pliers [3]
    Knife [4]
    Vinyl plastic electrical tape

1.   Unplug the cord. Check the cord for any signs of damage.

If the plug on end of cord is damaged, go to Page 33 to replace the plug.

If the cord or plug is not damaged, the problem is in the switch or socket. Perform the following procedures to repair the lamp socket or replace the cord.

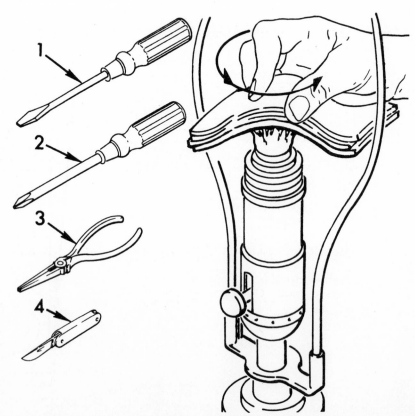

Repairing Lamps

Lamp shade is usually attached to lamp by a decorative nut or stud.

2. Remove lamp shade. Remove bulb.

3. Using knife, carefully pry off protective felt pad [1].

If installed:

4. Remove nut [5] and weight [4]. Lift tube [9] up about 6 inches. Turn tube counterclockwise or loosen set screw [8] to remove tube from socket.

5. Press in at bottom of outer shell [2] and lift straight up. Remove insulated shell [3].

6. Loosen two screws [6]. Remove wires [7].

You should take old socket with you when buying replacement.

7. Loosen screws [6]. Install wires [7] on new socket. Position wires around screws so that as screws are tightened, loops [10] in ends of wires are closed.

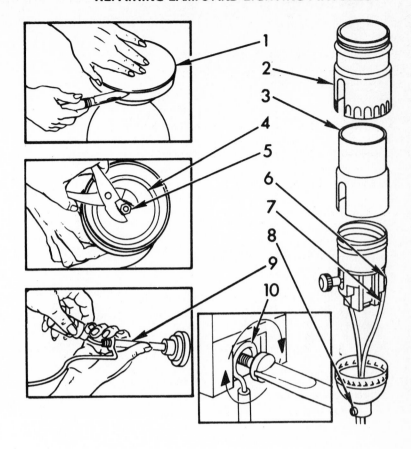

Repairing Lamps

8. Tighten loop around screws [7] with long nose pliers. Tighten screws.

9. Pull on cord [11] until socket [6] fits firmly into cap [9]. Install insulated shell [3].

10. Place outer shell [1] over socket [6]. Position shell at a slight angle. Press bottom of shell into cap [9].

If removed:

11. Install socket on tube [2]. Tighten screw [10]. Install weight [4] and secure with nut [5]. Install felt pad [8].

12. Install bulb. Install lamp shade. Install decorative nut or stud.

13. Plug in cord. Check operation of lamp.

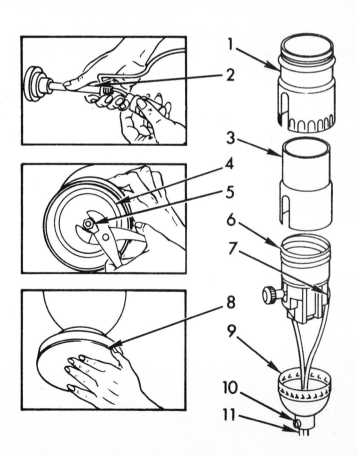

REPAIRING LAMPS AND LIGHTING FIXTURES

▶ **Repairing Lighting Fixtures**

There are many different sizes and types of wall
and ceiling fixtures. They are mounted in several
different ways.
The best procedure to follow when repairing a
fixture is to pay strict attention to the order of
disassembly. This will aid you when reassembling
it. If a lighting fixture goes bad, it is usually the
socket that needs replacement.

Procedures for repairs are provided as follows:
    Incandescent Lighting:
      Ceiling Fixtures, see section below.
      Wall Fixtures, Page 53.
    Fluorescent Lighting, Page 54.

▶ **Repairing Ceiling Fixtures**

The following tools are required:
    Common pliers [1]
    Screwdriver, common [2] or Phillips [3]
    Diagonal cutting pliers [4]
    Knife [5]

WARNING

**Be sure circuit breaker is OFF or fuse removed for
circuit being worked on.**

1.   Remove decorative covering.

2.   Check bulb. Replace, if required.

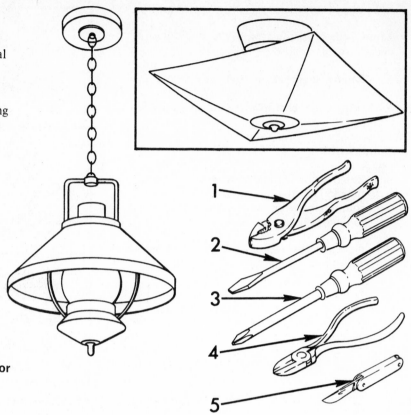

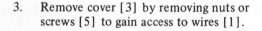

3.   Remove cover [3] by removing nuts or
    screws [5] to gain access to wires [1].

4.   Check wire connections. Tighten as required.

If connections were loose, check operation of
light. If light still does not work, continue.

5.   Remove connectors [2] by turning counter-
    clockwise. Remove fixture.

Wires [4] may be permanently attached to
socket [6] or held to socket with screws.

6.   Remove socket [6].

You should take the old socket with you when
buying a replacement.

If new socket has wires attached, cut and trim
insulation from new wires. Follow procedures
on Page 26. Then go to Page 53, Step 9.

If new socket does not have wires attached,
go to Page 53, Step 7.

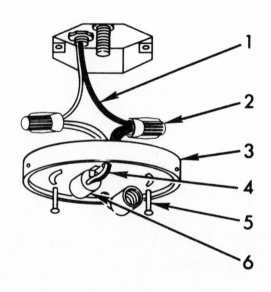

7. Install wires [3] on socket [5]. Position wire around screw [6] so that tightening screw closes loop [7].

8. Tighten loop [7] around screw [6] with long nose pliers. Tighten screws.

9. Thread wires through fixture, if required.

10. Install socket [5].

Most wires are color-coded white or black. If there is not a white wire, the wire with a colored tracer line connects to the white wire. (For an illustration of a tracer line, see Note following Step 12, Page 54.) Connect black to black; white to white.

11. Connect wires [3] to outlet box wires [1]. Follow procedures on Page 28 (top).

12. Install cover [2] by installing screws [4] or nuts.

13. Install bulb(s).

14. Turn circuit breaker to ON or install fuse. Check operation of light.

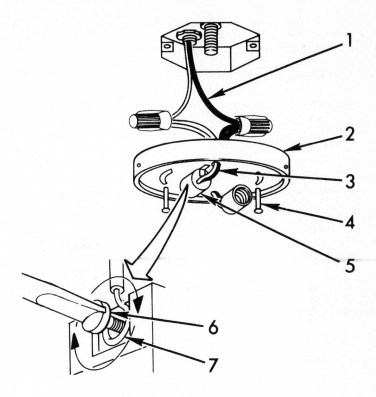

▶ Repairing Wall Fixtures

The following tools are required:

> Screwdriver, common [1] or Phillips [2]
> Long nose pliers [3]
> Knife [4]
> Diagonal cutting pliers [5]

WARNING

Be sure to turn circuit breaker off or remove fuse that controls circuit being worked on.

1. Check bulb. Replace, if required.

2. Remove lamp shade. Remove bulb.

Wall fixtures [6] are usually held to the wall by one or two decorative thumb nuts [7].

3. While holding fixture [6], remove thumb nuts [7]. Allow fixture to hang by wires [8].

4. Check wire connections [9]. Tighten as required.

If connections were loose, check operation of light. If light still does not work, continue.

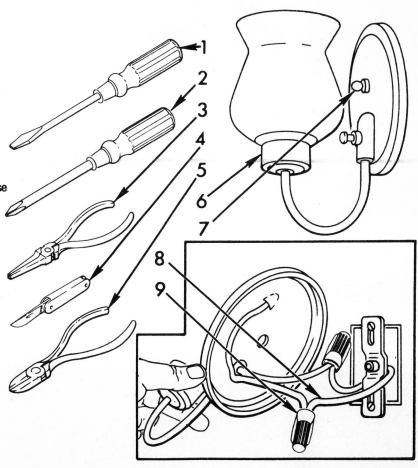

REPAIRING LAMPS AND LIGHTING FIXTURES

Repairing Wall Fixtures

5. While holding fixture [1], remove connectors [4, 5] by turning counterclockwise. Remove fixture.

If fixture has switch [3] installed, it is bad if it has to be positioned "just so" to turn on the light, or if the pull chain sticks when operated.

If the switch is not bad, socket is bad. Go to Step 8. Read Note before Step.

If switch is bad, continue.

6. Remove nut [2]. Remove switch [3]. Take switch with you when buying a new one to insure exact replacement.

7. Install switch [3]. Install nut [2]. Go to Step 13.

Wires [8] may be permanently connected to socket [6]. Cut wires off, if required.

8. Loosen screws [7]. Remove wires [8]. Remove socket by loosening screw [9].

You should take the old socket with you when buying replacement.

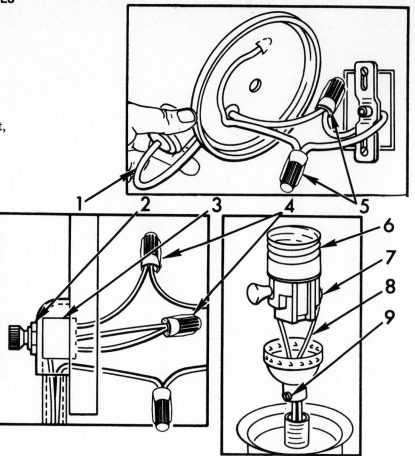

Repairing Wall Fixtures

9. Using old wires [6] as a guide, cut and trim insulation from wires. Follow procedures on Page 26.

If wires are already attached to new socket, go to Step 12.

10. Install wires [6] on new socket [3]. Position wires around screws [4] so that as screws are tightened, loops [2] in ends of wires are closed.

11. Tighten loops [2] around screws [4] with long nose pliers. Tighten screws.

12. Install socket [3]. Tighten screws [4].

Most fixture wires are color-coded — black and white. If there is not a white wire, the wire with a colored tracer line [1] connects to the white wire.

13. Connect fixture wires [6] to outlet box [9] with connectors [8]. Follow procedures on Page 28 (top).

14. Install fixture [5]. Install thumb nut [7].

15. Install bulb. Install shade.

16. Turn circuit breaker to ON or install fuse. Check operation of lamp.

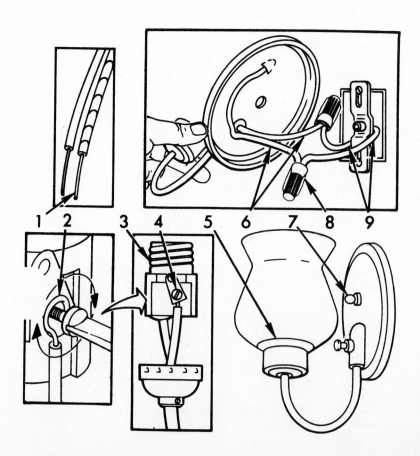

▶ **Repairing Fluorescent Lighting Fixtures**

The popularity of fluorescent lighting is steadily increasing. The main reason for this is that a fluorescent light gives off several times more light, for amount of wattage consumed, than an incandescent light. The tube will last from 5 to 10 times longer. Fluorescent lights also produce less heat and less glare than incandescent lights.

There are two types — the tubes [4] and the circline [7]. They are made with either a replaceable starter unit [5] or a rapid start. A rapid start is a starter built into the ballast [6].

Whenever replacing starters [5] or ballasts [6], be sure they are the same type and size as the ones removed.

First determine what is wrong with the fixture by using the chart below. Then go to the appropriate Step to fix the problem. Start with Number 1 and continue until problem is solved.

The following tools are required:

    Screwdriver, common [1] or Phillips [2]
    Long nose pliers [3]

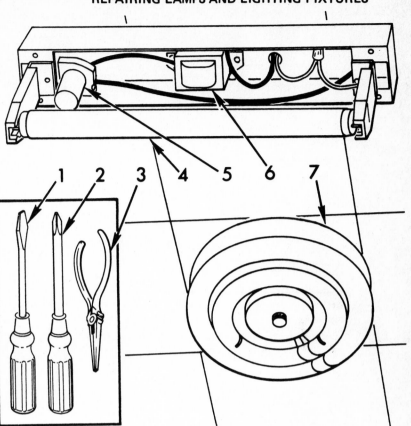

Fluorescent Lighting Checklist

Tube will not light	1.	Check that fuse or circuit breaker is good (Pages 24 and 25)
	2.	Replace starter (Step 5)
	3.	Replace tube (Step 1 or 3)
	4.	Replace ballast (Step 7)
Light blinks on and off	1.	Be sure tube is seated in socket (Step 1 or 3)
	2.	Remove tube. Straighten and lightly sand contacts on tube and in socket. Install tube. (Step 1 or 3)
	3.	Replace starter (Step 5)
	4.	Replace ballast (Step 7)
Light flickers and swirls inside tube	1.	If tube is new, this condition is normal. It will become steady with use.
	2.	If it does continue, replace starter (Step 5)
Humming or buzzing sound	1.	Check that ballast wire connections are tight.
	2.	Replace ballast with low-noise ballast.
Discolored tube	1.	Slightly brown color is normal
	2.	If black, and tube is newly installed, replace starter (Step 5)
	3.	If black, and tube is not new, replace the tube (Step 1 or 3)
	4.	If tube is discolored on one side only, remove it and turn it over (Step 1 or 3)
	5.	If tube is discolored on one end only, remove it and reverse ends (Step 1 or 3)

WARNING

Be sure to turn off circuit breaker or remove fuse that controls circuit being worked on.

REPAIRING LAMPS AND LIGHTING FIXTURES

Repairing Fluorescent Lighting Fixtures

WARNING

Be sure to turn off circuit breaker or remove fuse that controls circuit being worked on.

You may have to remove cover plate for access to fixture components. It may be fastened with decorative thumb nuts.

Replace Tube

Tube Type

1. Turn tube [5] 1/4 turn and remove.
2. Place tube [5] in slot [3]. Turn tube 1/4 turn. Go to Step 13.

Circline Type

3. Carefully disconnect tube [9] from socket [8]. Pull tube from retaining clips [4].
4. Connect new tube [9] to socket [8]. Press tube into retaining clips [4]. Go to Step 13.

Replace Starter

5. Turn starter [2 or 7] 1/4 turn counterclockwise and remove.
6. Place starter [2 or 7] in slot [1 or 6]. Turn starter 1/4 turn clockwise. Go to Step 13.

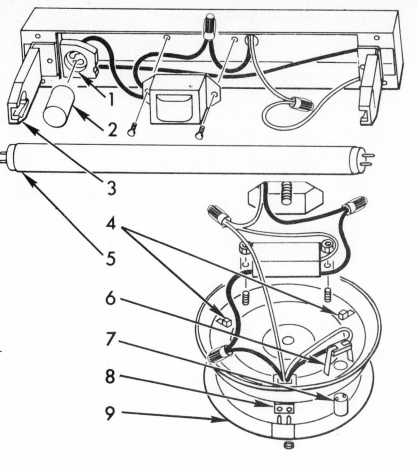

Replace Ballast

7. Remove tube (Step 1 or 3).

8. Label and disconnect all ballast wires [2 or 5] from fixture by removing connectors [1 or 6] or loosening screws.

9. While holding ballast [4 or 8], remove mounting screws [7] or nuts [3]. Remove ballast.

10. While holding new ballast [4 or 8] in place, install mounting screws [7] or nuts [3].

11. Connect all ballast wires [2 or 5] to fixture by installing connectors [1 or 6] or tightening screws. Follow procedures on Page 28 (top) for installing connectors. Remove labels.

12. Install tube (Step 2 or 4).

13. Install cover plate, if removed.

14. Turn circuit breaker to ON or install fuse. Check operation of light.

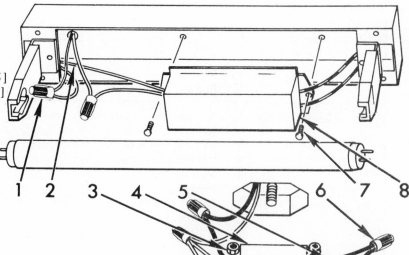

Doorbells, chimes and buzzers all work the same way and are wired in the same manner. These systems use low-voltage wiring, so that power need not be turned off when working with the wiring or components other than the transformer [2].

Most all systems today use a transformer [2] — not batteries. Batteries go bad quickly and are expensive. Transformers rarely go bad and require no maintenance.

If the noise mechanism (bell, chime or buzzer) [1] goes bad, it is usually best to replace it with a new one rather than try to repair it.

The other two problems can be:

- Wiring. Because wires are small (size 18 usually), they have a tendency to fray and break easily.

- Push buttons [3]. Because they are exposed to weather and use, the contacts become bent or corroded.

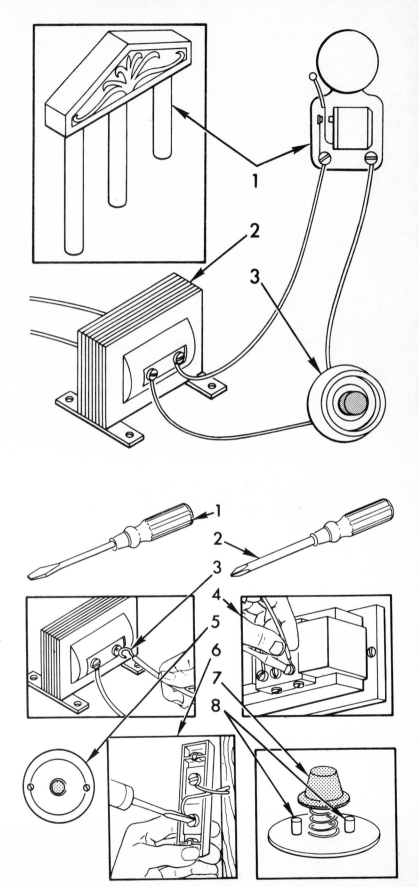

The following tools and supplies may be required to repair doorbell, chime or buzzer systems:

    Screwdriver, common [1] or Phillips [2]
    Vinyl plastic electrical tape
    Fine sandpaper or emery cloth

WARNING

If working on transformer, be sure to turn off main power switch.

1.   Check that all wire connections are tight and clean at transformer [3]. Tighten as required.

2.   Remove cover from noise mechanism [4]. Check that wire connections are tight and clean. Tighten as required. Install cover.

3.   Remove push-button plate [5]. Check that wire connections [6] are tight and clean. Tighten as required.

4.   Inspect wiring throughout system. Tape frayed or bare wires as required.

5.   Carefully pull out push button [7] for access to contacts [8].

6.   Using sandpaper or emery cloth, lightly sand contacts.

7. Push the button [1]. Check that contacts [2] touch. Carefully bend as required.

If contacts [2] are broken or bent beyond repair, replace the push-button assembly.

8. Install plate [3].

9. While someone pushes the button [1] check the transformer [4] for following indications:

- If bell rings, your problem has been solved.

- If bell doesn't ring and there is a hum in the transformer [4], the noise mechanism [5] is bad.

- If bell doesn't ring and there is no hum in transformer [4], the transformer is bad.

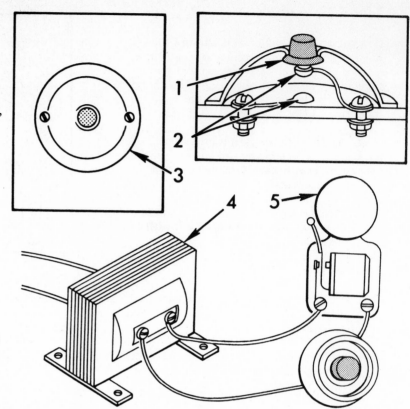

INSTALLING FIXTURES and WIRING

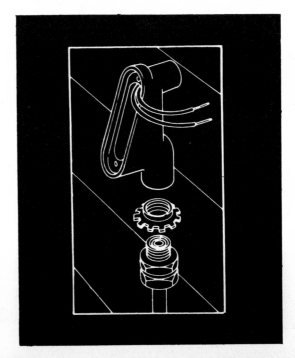

It is a simple process to install wiring and fixtures before a house is completely built. However, once built, extending existing wiring for new fixtures becomes more difficult. Often it requires extensive carpentry.

The procedures in this section describe installation of indoor and outdoor fixtures and wiring and installation of surface wiring.

Before performing any procedures in this section, you should read the first section of this book to become familiar with the electrical system.

When undertaking a new wiring job, proceed as follows:

1. Plan — Plan exactly what you want to do. Then check with your local codes to see if it is permitted.

    Carefully measure and sketch the route and position of new wiring and fixtures. Be sure that your new installation will not overload an existing circuit.

2. Identify all tools and supplies — Be sure you have all the tools you need before beginning.

This will save you unnecessary trips to the store in the middle of the job.

Buy supplies according to your sketch. You may want to take the sketch to your dealer. He can help you with the types and quantities of supplies needed. Whenever adding new wiring, be sure to use a flexible cable. Usually armored BX cable or non-metallic sheathed cable is best for indoors. Outdoor wiring depends on local code requirements.

3. Install fixtures and wiring — After planning the job and obtaining all required tools and supplies, you are ready to begin. Read the following pages before starting any procedure to become familiar with techniques and operations which will be required:

    ●   How to Use a Fish Tape, Page 85.
    ●   How to Connect Nonmetallic Sheathed Cable, Page 85.
    ●   How to Connect Armored BX Cable, Page 86.

■ INSTALLING INDOOR FIXTURES AND WIRING ■

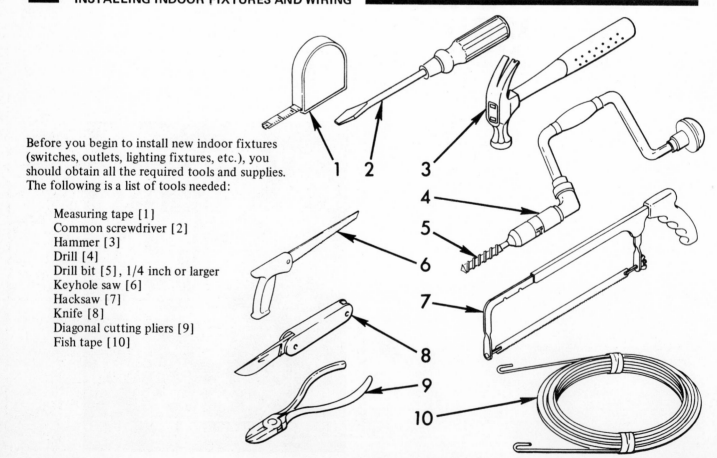

Before you begin to install new indoor fixtures (switches, outlets, lighting fixtures, etc.), you should obtain all the required tools and supplies. The following is a list of tools needed:

Measuring tape [1]
Common screwdriver [2]
Hammer [3]
Drill [4]
Drill bit [5], 1/4 inch or larger
Keyhole saw [6]
Hacksaw [7]
Knife [8]
Diagonal cutting pliers [9]
Fish tape [10]

The following supplies may also be needed, depending on the requirements of your specific job:

Flexible cable. Armored BX [1] or nonmetallic sheathed cable [2] is recommended.

Cable connectors. Armored BX connectors [3] or nonmetallic sheathed cable connectors [4]. Fiber bushings [5] are required with armored BX connectors.

Electrical boxes. Many types, sizes and mounting methods are available. Ask your dealer what type is best for your needs:

Electrical box [6] is used for mounting switches and outlets. For a description of how to attach this type of box to the wall, see bottom half of this page.

Octagon electrical box [8] is used for mounting lighting fixtures or for use as a junction box. For ceiling fixtures, a bar hanger [7] may be required to attach the box.

Side-bracket electrical box [9] is used for mounting switches and outlets. It is used for new construction only where bracket [10] can be attached directly to studs.

Insulated solderless connectors [11]. Use quality connectors.

Fixture. Be sure the fixture you purchase fits correctly in electrical box.

The method of mounting electrical boxes [2] depends on the type of walls you have.

Lath and plaster walls require attaching the box directly to the lath [4].

Drywall construction requires attaching the box to either a stud [1] or to the drywall [3] itself, depending on the type of box you purchase.

Most electrical boxes can be mounted to both lath and plaster walls and drywall by positioning the mounting ears [5] as shown for each type of wall.

When you purchase an electrical box, be sure that the type you buy is suited for the type of wall construction in your house.

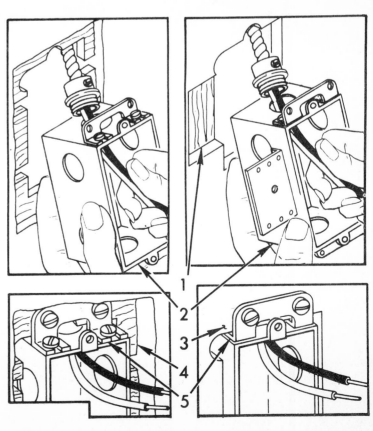

INSTALLING INDOOR FIXTURES AND WIRING

To install a new fixture, you will have to cut an opening in the wall for the electrical box and fixture.

The location of the opening depends on the mounting method of the electrical box, the type of fixture to be installed, and the method you use to extend the wiring from the power source.

Be sure of the position of wall studs before cutting the opening. An accurate cut will prevent unnecessary wall repairs after your new fixture is installed.

Wall switches [1] should be located with their centers approximately 4 feet above the floor.

Outlets [2] are located with their centers approximately 12 inches above the floor.

Countertop switches and outlets [3] are located with their centers approximately 8 inches above the countertop.

The dimensions for locating switches and outlets are given as a general guide only. For a good appearance, you will want to mount new fixtures at the same height as existing fixtures in the house.

The following procedures for extending wiring from an existing power source to the location of a new fixture, and for installing the fixture itself, are included in this section. Try to select the simplest and most direct method possible for your particular wiring needs:

- Through an attic [1]
- Behind baseboards [2]
- Around a corner [3]
- Through a wall [4]
- Through a basement [5]

Extending wiring through an attic [1] is perhaps the best method of getting power from a source on one wall to an opposite or adjacent wall or to a location on the ceiling. These procedures begin on Page 63.

Extending wiring behind a baseboard [2] is an easy method of getting power from a source on the same wall as the new fixture. However, if a doorway obstructs the baseboard between the power source and new fixture, an alternate method of extending wiring should be selected. Procedures for extending wiring behind baseboards begin on Page 66.

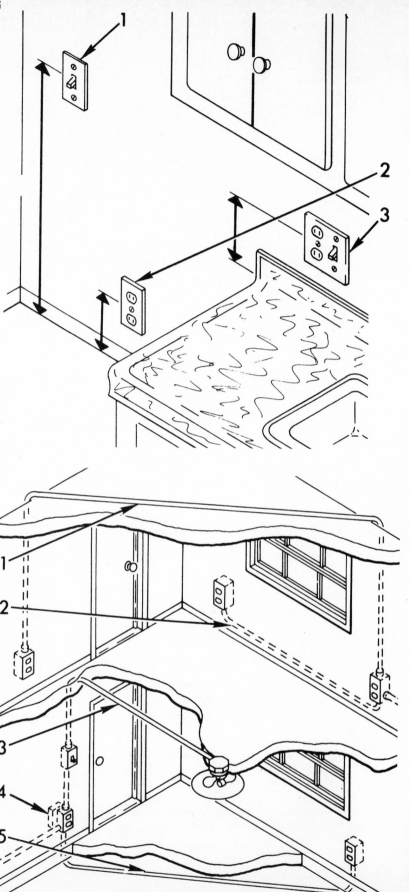

Extending wiring around a corner [1] between a wall and ceiling is probably the most difficult and time-consuming method. It is used to install a ceiling fixture when no access above the ceiling is available. The location of the power source must be directly in line with the new fixture location. The cable run must be completely unobstructed by ceiling joists. These procedures begin on Page 68.

Extending wiring directly through a wall [2] is the easiest method if the new fixture installation is directly opposite the power source. The location of the power source and new fixture must be unobstructed by wall studs. Otherwise, another method must be selected. Extending wiring through a wall begins on Page 71.

Extending wiring through a basement [3] is a good method to use if the basement ceiling joists are exposed to provide access. The fixture to be used as the power source can be any existing fixture near the new location. These procedures begin on Page 73.

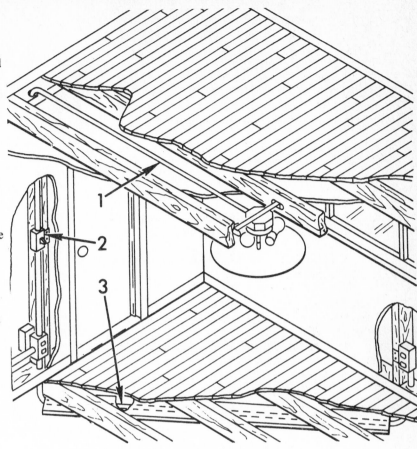

▶ **Extending Wiring Through an Attic**

Read through entire procedure before beginning.

<u>WARNING</u>

Be sure to turn off circuit breaker or remove fuse that controls circuit being worked on.

1.   From your plans, locate power source A for new installation D. Remove cover plate [2] and pull out fixture [3] to gain access to electrical box.

Knockout [1] to be removed should face the direction of the wire extension. Use the top knockout, if available, for extending to the attic.

2.   Using screwdriver, punch out knockout [1] from electrical box.

3.   At location of new fixture D, locate studs by tapping on the wall or using a magnetic stud finder.

Location of opening for new fixture D depends on how the electrical box is to be attached and the type of fixture to be installed. Read information on Pages 61 and 62 before continuing.

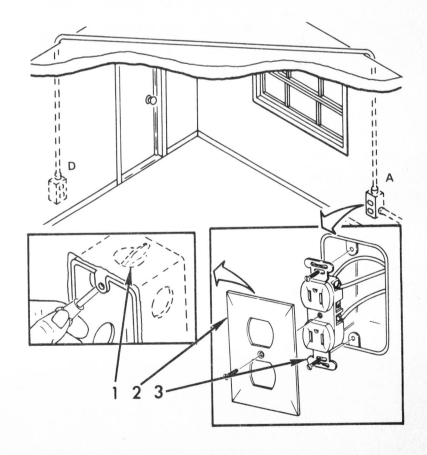

Extending Wiring Through an Attic

If walls are of drywall construction, go to Step 5.

If walls are of lath and plaster construction, continue.

4. Drill hole [1] at approximate center of location of new fixture D. Enlarge hole enough to locate laths.

When drawing outline of electrical box, leave enough lath at top and bottom of outline to attach electrical box.

5. Draw outline of electrical box on wall at desired location of new fixture D. Use template [3] or new electrical box as a guide.

6. Drill starter holes [2] for keyhole saw blade. Cut around outline of electrical box. Remove cutout [4].

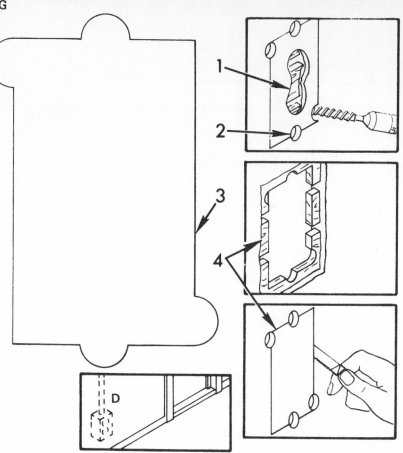

Extending Wiring Through an Attic

You may have to drill at an angle to avoid obstructions.

7. Working from attic, drill hole B directly above source A. Drill hole C directly above new fixture hole D. Enlarge holes with keyhole saw until cable and connectors will pass through easily.

8. Remove 6 inches of outer insulation from end of cable [1].

9. Place cable connector [3] over wires and onto cable [1]. Tighten screws [2].

10. Fish cable [1] from hole B through electrical box knockout at source A.

11. Connect cable [1] to electrical box [4].

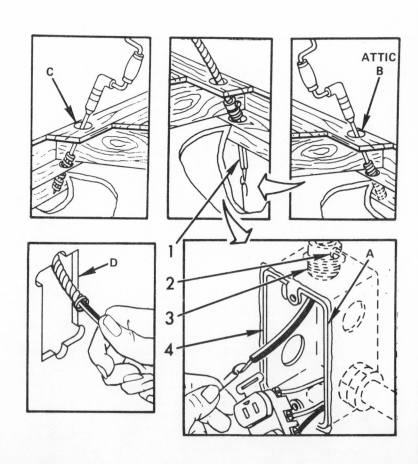

Extending Wiring Through an Attic

12. Cut cable [1] to approximate length required to extend through opening D. Remove 6 inches of outer insulation from end of cable.

13. Fish cable [1] through hole C to opening D.

14. Connect cable [1] to new electrical box [2].

15. Attach electrical box [2] to wall with screws or nails.

Cable must be secured to ceiling joists. Joists should be notched and cable laid into notches, then secured with insulated staples. Check local code for information on securing cables.

16. Secure cables to ceiling joists.

17. Trim off about 1/2 inch of insulation from wire ends [3] at electrical box [2].

18. Connect wires [3] to new fixture [5]. See Page 26 for correct wire connections.

19. Install fixture [5] by installing mounting screws [4]. Install cover plate [6].

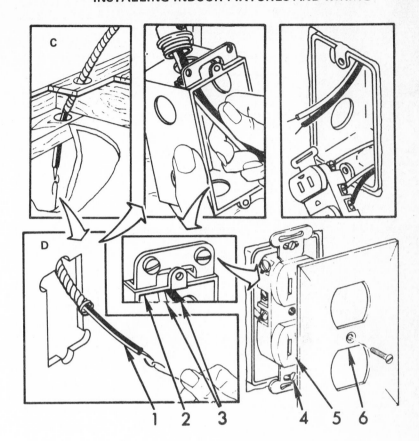

Extending Wiring Through an Attic

20. Trim off about 1/2 inch of insulation from wire ends [3] at source A.

21. Connect wires [3] to source wires [4], black to black and white to white.

22. Connect bare ground wire [6] to electrical box screw [7] or ground connector [5], if required.

23. Install fixture [2] and cover plate [1].

24. Turn circuit breaker to ON or install fuse. Check operation of new fixture.

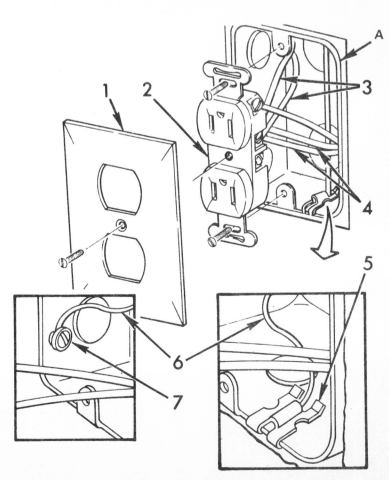

INSTALLING INDOOR FIXTURES AND WIRING

▶ **Extending Wiring Behind Baseboards**

Read through entire procedure before beginning.

WARNING

Be sure to turn off circuit breaker or remove fuse that controls circuit being worked on.

1. From your plans, locate power source A for new installation D. Remove cover plate [1] and pull out any fixture [2] to gain access to electrical box.

Knockout [3] to be removed should face the direction of the wire extension. Use the bottom knockout, if available, for a direct angle to the baseboard.

2. Using screwdriver, punch out knockout [3] from electrical box.

3. At location of new fixture D, locate studs by tapping on the wall or using a magnetic stud finder.

Location of opening for new fixture D depends on how the electrical box is to be attached and the type of fixture to be installed. Read information on Pages 61 and 62 before continuing.

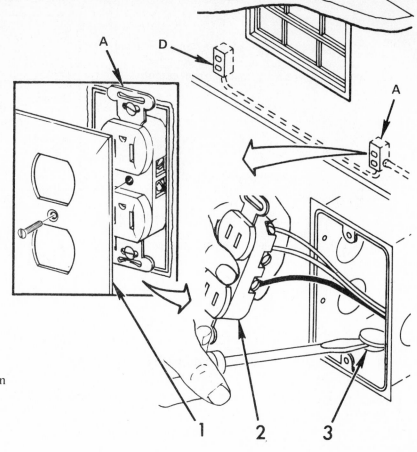

Extending Wiring Behind Baseboards

If walls are of drywall construction, go to Step 5.

If walls are of lath and plaster, continue.

4. Drill hole [1] at approximate center of location of new fixture D. Enlarge hole enough to locate laths.

When drawing outline of electrical box, leave enough lath at top and bottom of outline to attach electrical box.

5. Draw outline of electrical box on wall at desired location of new fixture D. Use template [3] or new electrical box as a guide.

6. Drill starter holes [2] for keyhole saw blade. Cut around outline of electrical box. Remove cutout [4].

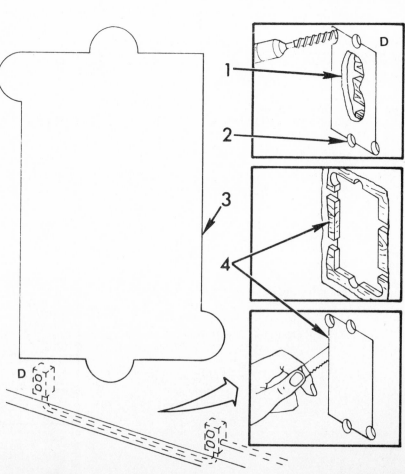

Extending Wiring Behind Baseboards

7. Measure distance B to C along baseboard from source of power A to new fixture installation D.

8. Remove length of baseboard. Cut two holes B and C behind baseboard.

9. Cut a groove from hole B to hole C. Be sure it is deep enough to receive cable when baseboard is installed.

10. Remove 6 inches of outer insulation from end of cable [1].

11. Place cable connector [3] over wires and onto cable [1]. Tighten screws [2].

12. Fish cable [1] from hole B through electrical box knockout at source A.

13. Connect cable [1] to electrical box [4].

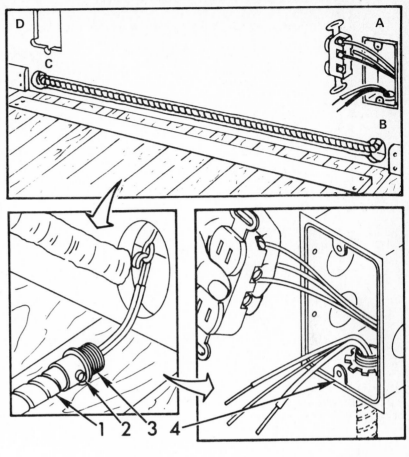

Extending Wiring Behind Baseboards

14. Cut cable [6] to approximate length required to extend through opening D. Remove 6 inches of outer insulation from end of cable.

15. Fish cable [6] through hole C to opening D. Place cable neatly in groove [7].

16. Connect cable [6] to new electrical box [1].

17. Attach electrical box [1] to wall with screws or nails.

18. Trim off about 1/2 inch of insulation from wire ends [2] at electrical box [1].

19. Connect wires [2] to new fixture [3]. See Page 26 for correct wire connections.

20. Install fixture [3] by installing mounting screws [4]. Install cover plate [5].

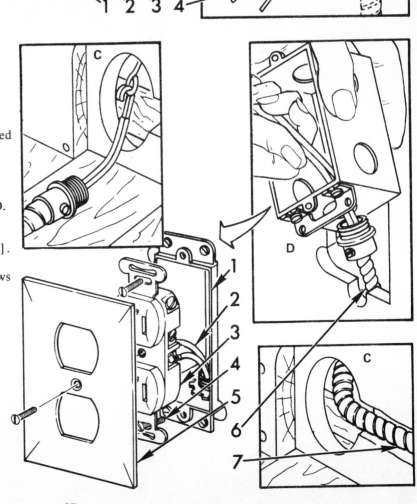

Extending Wiring Behind Baseboards

21. Trim off about 1/2 inch insulation from wire ends [7] at source A.

22. Connect wires [7] to source wires [6], black to black and white to white.

23. Connect bare ground wire [2] to electrical box screw [1] or ground connector [3] if required.

24. Install fixture [5] and cover plate [4].

25. Turn circuit breaker to ON or install fuse. Check operation of new fixture.

26. Install baseboard.

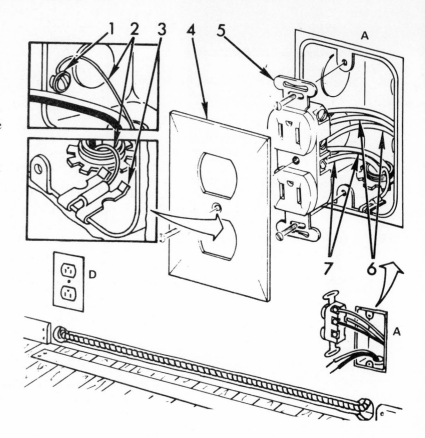

▶ **Extending Wiring Around a Corner**

Read through entire procedure before beginning.

WARNING

Be sure to turn off circuit breaker or remove fuse that controls circuit being worked on.

A 12-foot and a 25-foot fish tape are required to extend wiring around a corner.

1. From your plans, locate power source A for new installation C. Be sure source A is directly in line with new installation C, with the route for the cable completely unobstructed by studs.

2. Remove cover plate [4] and fixture [3]. Remove electrical box [2] from wall.

Knockout [1] to be removed should face the direction of the wire extension. Use the top knockout, if available, for a direct angle to the ceiling.

3. Using screwdriver, punch out knockout [1] from electrical box [2].

4. At location of new fixture C, locate joists by tapping on ceiling or using a magnetic stud finder.

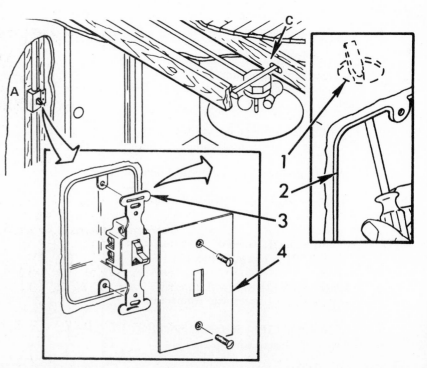

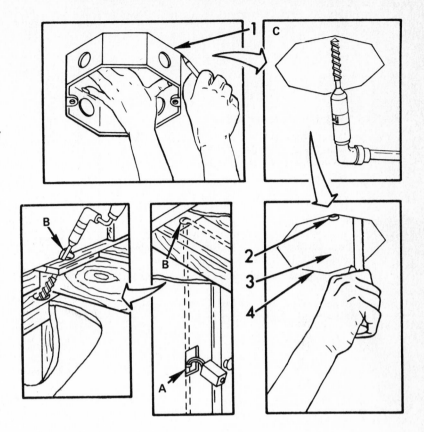

Extending Wiring Around a Corner

5. Using new electrical box [1] as a guide, draw outline of box at desired location of new fixture C.

6. Drill starter hole [2] for keyhole saw blade. Cut around outline of electrical box [4]. Remove cutout [3].

Hole B must be drilled from second floor or attic directly above source A. If working from second floor, try to remove a baseboard for hole B to conceal the hole.

7. Working from second floor or attic, drill hole B down toward source A.

Extending Wiring Around a Corner

8. Bend both ends of 12-foot fish tape [1] into hooks.

9. Push one end of fish tape [1] into hole B and down toward source A. Pull end through opening at source A, leaving other end at hole B.

10. Bend both ends of 25-foot fish tape [2] into hooks.

11. Push one end of fish tape [2] into opening C until 12-foot fish tape [1] is touched.

12. Carefully pull on either fish tape [1 or 2] until both tapes are hooked together.

13. Slowly pull on 12-foot fish tape [1] at source A until 25-foot fish tape [2] extends from opening C through source A. Disconnect fish tapes.

14. Remove 6 inches of outer insulation from cable [3].

15. Using 25-foot fish tape [4] fish cable [3] from opening C through source A. Disconnect fish tape.

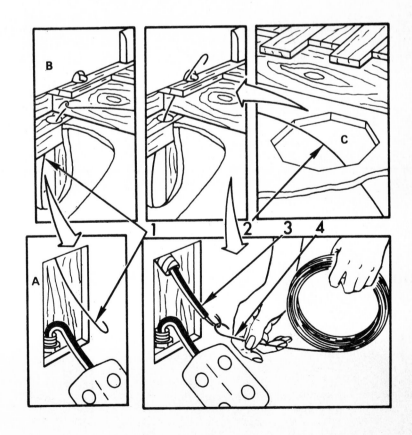

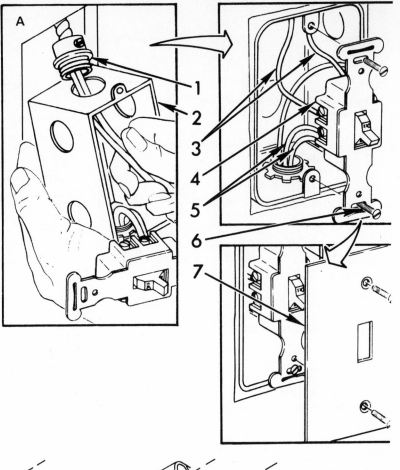

Extending Wiring Around a Corner

16. At source A, connect cable [1] to electrical box [2]. Attach electrical box to wall with screws or nails.

17. Trim off about 1/2 inch of insulation from wire ends [3].

18. Connect wires [3] to fixture [4]. Connect wires [5] to fixture. See Page 26 for correct wire connections.

19. Install fixture [4] by installing mounting screws [6]. Install cover plate [7].

Extending Wiring Around a Corner

20. Connect cable [2] to electrical box [3].

Electrical box [3] may require attaching to the ceiling by either screws or nails or by a bar hanger [1]. Hanger attaches through rear knock-out of box and is secured by nut [4].

21. Attach electrical box [3] to ceiling.

22. Trim off about 1/2 inch of insulation from wire ends [5].

Wire connection between wires [5] and fixture may require connectors [6] or screw terminals, depending on type of fixture being installed. Attach black wire to black, and white to white.

Attaching hardware required by fixture varies. Follow manufacturer's instruction for attaching fixture to ceiling.

23. Turn circuit breaker to ON or install fuse. Check operation of fixture.

24. Install baseboard on second floor, if required.

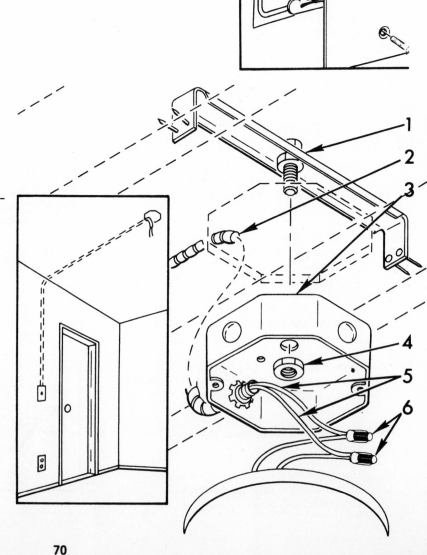

▶ **Extending Wiring Through a Wall**

Read through entire procedure before beginning.

WARNING

Be sure to turn off circuit breaker or remove fuse that controls circuit being worked on.

1.   From your plans, locate power source A for new installation B. Be sure new installation B is directly opposite source A.

2.   Remove cover plate [3] and pull out fixture [2] to gain access to electrical box.

Knockout [1] to be removed should face the direction of the wire extension. Use the rear knockout, if available, for a direct angle to new installation B.

3.   Using screwdriver, punch out knockout [1] from electrical box.

4.   At location of new fixture B, locate studs by tapping on the wall.

Location of opening for new fixture B depends on how the electrical box is to be installed. Read information on Pages 61 and 62 before continuing.

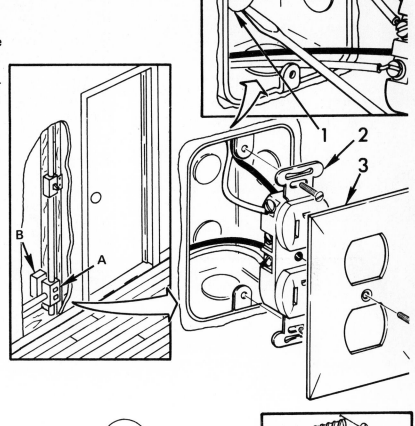

Extending Wiring Through a Wall

If walls are of drywall construction, go to Step 6.

If walls are of lath and plaster, continue.

5.   Drill hole [2] at approximate center of location of new fixture B. Enlarge hole enough to locate laths.

When drawing outline of electrical box, leave enough lath at top and bottom of outline to attach electrical box.

6.   Draw outline of electrical box on wall at desired location of new fixture B. Use template [3] or new electrical box as a guide.

7.   Drill starter holes [1] for keyhole saw blade. Cut around outline of electrical box. Remove cutout [4].

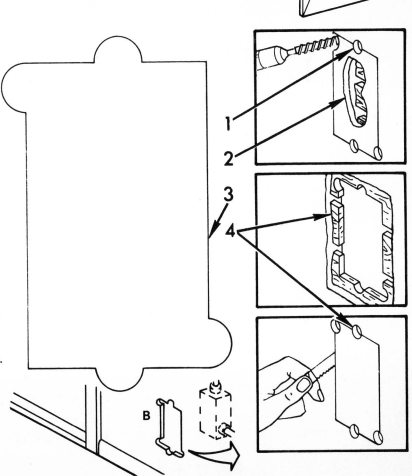

Extending Wiring Through a Wall

8. Remove 6 inches of outer insulation from end of cable [1]. Place cable connector [2] over wires and onto end of cable. Tighten screws [5].

9. Working at new fixture hole B, insert wires [6] through electrical box knockout at source A.

10. Connect cable [1] to electrical box [9].

11. Trim off about 1/2 inch of insulation from wire ends [6].

12. Connect wires [6] to source wires [8], black to black and white to white.

13. Connect bare ground wire [4] to electrical box screw [3] or ground connector [7], if required.

14. Install fixture [10] and cover plate [11].

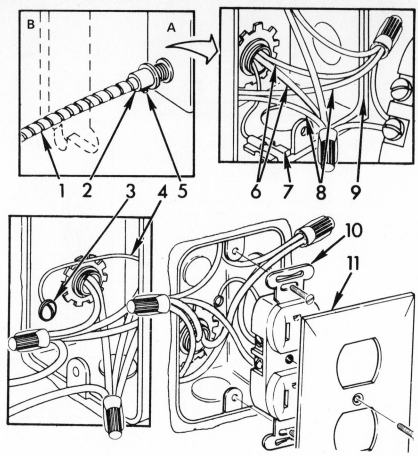

Extending Wiring Through a Wall

15. Cut cable [1] to approximate length required to extend through opening B. Remove 6 inches of outer insulation from end of cable.

16. Connect cable [1] to electrical box [2].

17. Attach electrical box [2] to wall with screws or nails.

18. Trim off about 1/2 inch of insulation from wire ends [3].

19. Connect wires [3] to new fixture [5]. See Page 26 for correct wire connections.

20. Install new fixture [5] by installing mounting screws [4]. Install cover plate [6].

21. Turn circuit breaker to ON or install fuse. Check operation of new fixture.

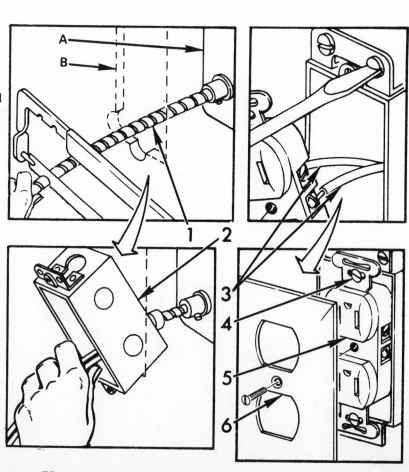

▶ **Extending Wiring Through a Basement**

Read through entire procedure before beginning.

WARNING

Be sure to turn off circuit breaker or remove fuse that controls circuit being worked on.

1. From your plans, locate power source A for new installation D. Remove cover plate [2] and pull out any fixture [1] to gain access to electrical box.

Knockout [3] to be removed should face the direction of the wire extension. Use the bottom knockout, if available, for a direct angle to the basement.

2. Using screwdriver, punch out knockout [3] from electrical box.

3. At location of new fixture D, locate studs by tapping on the wall or using a magnetic stud finder.

Location of opening for new fixture D depends on how the electrical box is to be attached and the type of fixture to be installed. Read information on Pages 61 and 62 before continuing.

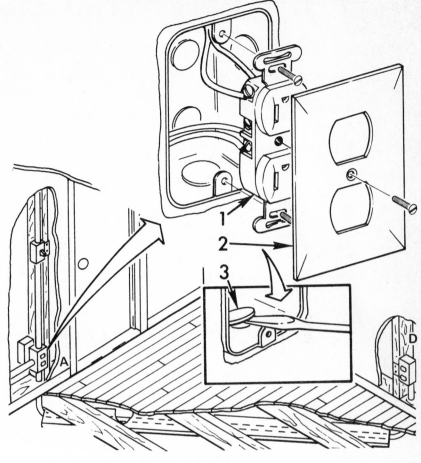

Extending Wiring Through a Basement

If walls are of drywall construction, go to Step 5.

If walls are of lath and plaster, continue.

4. Drill hole [2] at approximate center of location of new fixture D. Enlarge hole enough to locate laths.

When drawing outline of electrical box, leave enough lath at top and bottom of outline to attach electrical box.

5. Draw outline of electrical box on wall at desired location of new fixture D. Use template [3] or new electrical box as a guide.

6. Drill starter holes [1] for keyhole saw blade. Cut around outline of electrical box. Remove cutout [4].

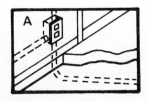

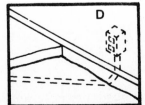

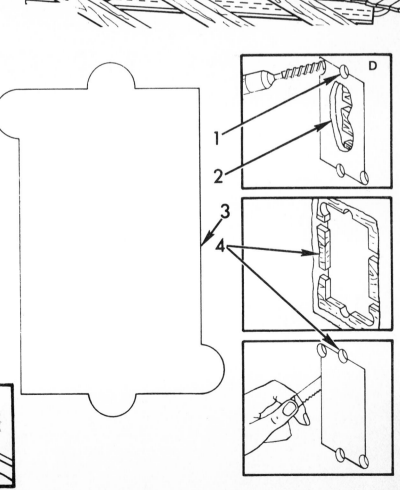

Extending Wiring Through a Basement

You may have to drill at an angle to avoid obstructions.

7. Working from basement, drill hole B directly under source A. Drill hole C directly under new fixture hole. Enlarge holes with keyhole saw until cable and connectors will pass through easily.

8. Remove 6 inches of outer insulation from end of cable [4].

9. Place cable connector [3] over wires and onto end of cable [4]. Tighten screws [2].

10. Fish cable [4] from hole B through electrical box knockout at source A.

11. Connect cable [4] to electrical box [1].

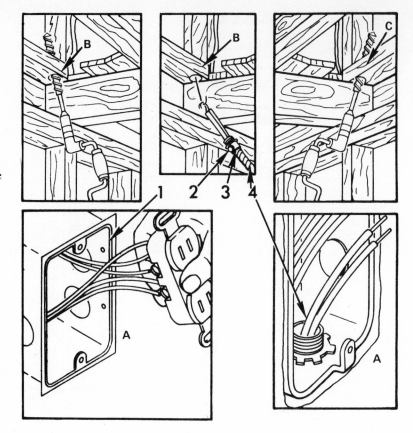

Extending Wiring Through a Basement

12. Cut cable [6] to approximate length required to extend through opening D. Remove 6 inches of insulation from end of cable.

13. Fish cable [6] through hole C to opening D.

14. Connect cable [6] to electrical box [5].

15. Attach electrical box [5] to wall with screws or nails.

Cable must be secured to basement ceiling. Check local code for information on securing cable.

16. Secure cable to basement ceiling.

17. Trim off about 1/2 inch of insulation from wire ends [1] at electrical box [5].

18. Connect wires [1] to new fixture [2]. See Page 26 for correct wire connections.

19. Install fixture [2] by installing mounting screws [3]. Install cover plate [4].

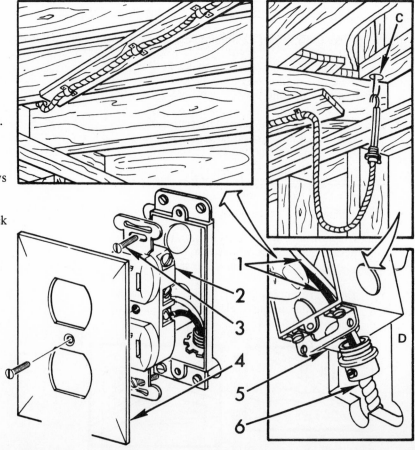

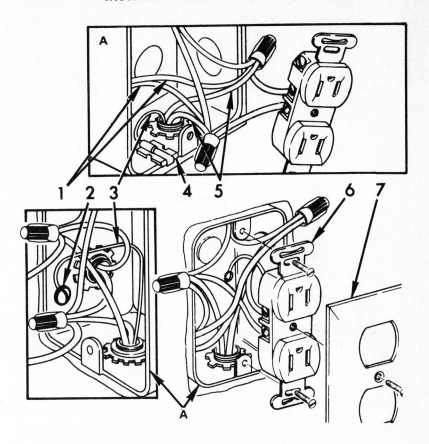

Extending Wiring Through a Basement

20. Trim off about 1/2 inch of insulation from wire ends [5] at source A.

21. Connect wires [5] to source wires [1], black to black and white to white.

22. Connect bare ground wire [3] to electrical box screw [2] or ground connector [4], if required.

23. Install fixture [6] and cover plate [7].

24. Turn circuit breaker to ON or install fuse. Check operation of new fixture.

INSTALLING OUTDOOR FIXTURES AND WIRING

Outdoor wiring differs from indoor wiring in that you must consider the effects of moisture and weather conditions. Therefore, special outdoor "rules" are mentioned by the Code.

In some communities, only conduit or lead-sheathed cable is permitted. In other communities, a type of plastic-covered cable is permissible, making it less expensive and simpler to install outdoor wiring. Before attempting an outdoor wiring project, be sure to check your local electrical code.

If installing outdoor wiring for swimming pool installations, leave the project to a professional. Nowhere is a person more subject to injury or death than in a pool in which the electrical wiring is installed improperly.

A ground fault circuit interrupter (GFCI) may be required for outdoor installations in your locality. Check specifically for this in your local code. Page 48 describes GFCI's.

Below are some important rules to follow when installing outdoor wiring:

- Use only weatherproof cable for outdoor wiring, even in conduit. Local codes specify the exact wiring for your area.

- Use conduit whenever wiring along a wall, when sharp rocks are present, or whenever there is a chance of the wires being cut or frayed.

- Never route wires through door or window openings where they may be damaged or frayed.

- Even though it is permissible in some areas to use wire nuts for outdoor electrical connections, it is recommended that all wire connections be soldered and taped with waterproof tape.

- Whenever working on outdoor wiring, even changing a bulb, be sure power to the circuit is turned off.

INSTALLING OUTDOOR FIXTURES AND WIRING

Installing outdoor fixtures requires extending wiring from an existing indoor power source to the location of your new outdoor fixture. The instructions in this section describe the following installations of outdoor fixtures (switches, outlets, lighting fixtures, etc.) and wiring:

- Installing a fixture [3] directly to an outside wall

- Extending wiring underground to a fixture [1] away from the house

Regardless of the type of fixture [1 or 3] you plan to install, an opening for the new fixture [3] or for the ell fitting [2] which begins the wire run away from the house must be made in an outside wall.

The location of the opening is generally a minimum of 18 inches above the ground. However, the height varies depending on local code requirements. Be sure to check your code before beginning. The location of the opening may affect which inside power source you use and the method of extending your wiring.

The power source [1] for your new installation should be as near as possible to the location of the opening in the outside wall. Plan the simplest and most direct route for the wiring from inside to outside.

For example, an outlet [1] directly opposite the planned exterior opening would be the most convenient point from which to obtain power. Wires [2] could then be extended directly through the wall, rather than require using a fish tape to extend the wiring a more complicated route. You may even want to extend your wiring indoors to provide a close source of power for the outdoor wiring.

Outdoor fixtures differ from indoor fixtures in that they are specifically designed to resist weather conditions. All outdoor fixtures provide a seal against dirt and moisture.

The actual switches and outlets [4] used in an outdoor fixture are the same as those used indoors. The weatherproof seal is provided by enclosed electrical boxes [3] and sealed cover plates [6] and gaskets [5].

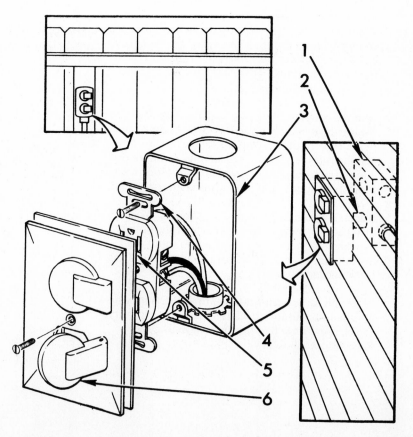

The procedures for extending wiring underground to a fixture [1] away from the house begin on Page 79. Procedures for installing a fixture [7] directly to an outside wall begin on this page.

▶ **Installing a Fixture on an Outside Wall**

The following tools and supplies are required:

Drill bit [2], 1/4 inch or larger (masonry bit if outside walls are plaster)
Drill [3]
Screwdriver [4]
Knife [5]
Fish tape [6]
Flexible cable [11], the length required to extend from your planned power source to the location of the fixture
Electrical box. Enclosed electrical box [10] is surface-mounted to the wall.
Electrical box [8] can be used when flush-mounted to the wall. Be sure electrical box [8] has provisions for mounting to your type of wall.
Outdoor fixture (switch or outlet) [9] with protective cover plate and gasket
Weatherproof caulking compound

Read through entire procedure before beginning.

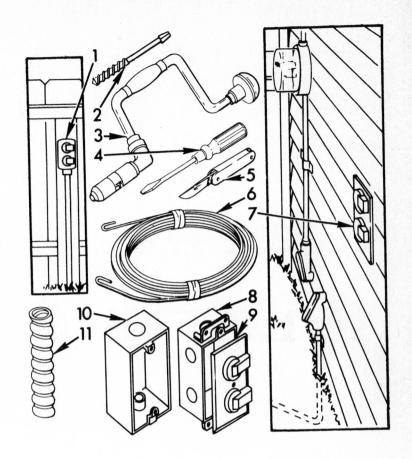

Installing a Fixture on an Outside Wall

WARNING

Be sure to turn off circuit breaker or remove fuse that controls circuit being worked on.

Wiring must be extended from an indoor power source [1] to an opening [2] in the outside wall. Instructions beginning on Page 62 describe different methods of extending wiring. Select one of the methods to determine the best power source to use and the location of the wall opening.

1. From your plans, locate power source [1] for your outdoor fixture. Remove cover plate [5] and pull out any fixture [4] to gain access to electrical box.

Knockout [3] to be removed should face the direction of the wire extension.

2. Using screwdriver, punch out knockout [3] from electrical box.

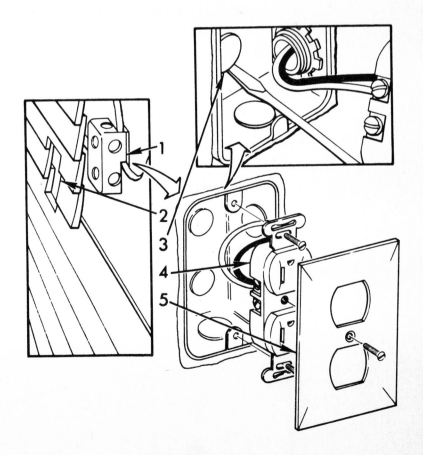

INSTALLING OUTDOOR FIXTURES AND WIRING

Installing a Fixture on an Outside Wall

Size of the opening [1] depends on the type of electrical box you plan to install. Flush-mounted box requires an opening the size of the box. Surface-mounted box requires an opening the size of the cable being connected.

3. On outside wall, make opening [1] the required size and location for your new fixture.

4. Extend wiring from power source [2] through opening [1]. Connect cable [3] to electrical box [6] of power source.

5. Connect wires [10] to source wires [8], black to black and white to white.

6. Connect bare ground wire [4] to electrical box screw [5] or ground connector [11] if required.

7. Install fixture [9] and cover plate [7].

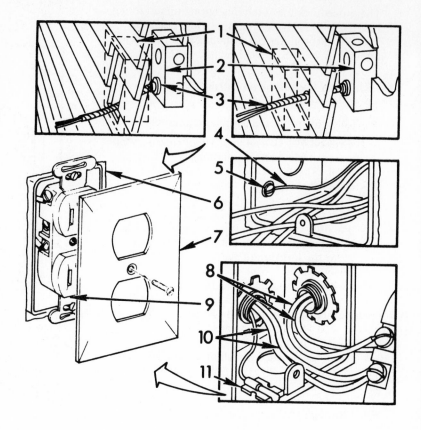

Installing a Fixture on an Outside Wall

8. At outside opening [1], cut cable [2] to required length.

9. Connect cable [2] to new electrical box [3].

10. Attach electrical box [3] to outside wall.

11. Connect wires [4] to new fixture [6]. See Page 26 for correct wire connections.

12. Install fixture [6] by installing mounting screws [5]. Install gasket [7] and cover plate [8].

13. Turn circuit breaker to ON or install fuse. Check operation of new fixture.

Weatherproof caulking compound is applied around surface-mounted electrical box [3] where it meets the wall. For flush-mounted fixtures, caulking compound can be applied around cover plate [8] where it meets the wall, if required.

14. Apply caulking compound around new fixture at the wall.

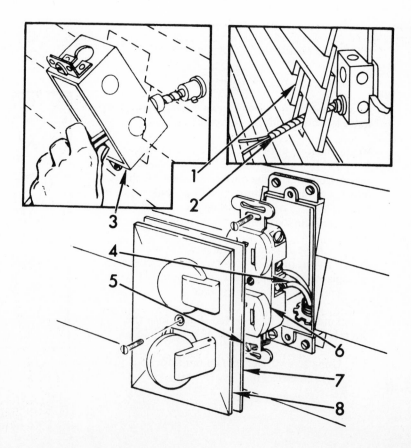

▶ Extending Wiring Underground

The following tools are required to extend wiring underground:

Drill [1]
Drill bit [2], 1/4 inch or larger
   (masonry bit if outside walls
   are plaster)
Screwdriver [3]
Knife [4]
Fish tape [5]
Adjustable wrench [6]
Hacksaw [7]
Reaming tool [8]
Conduit bender [9]
Shovel [10]

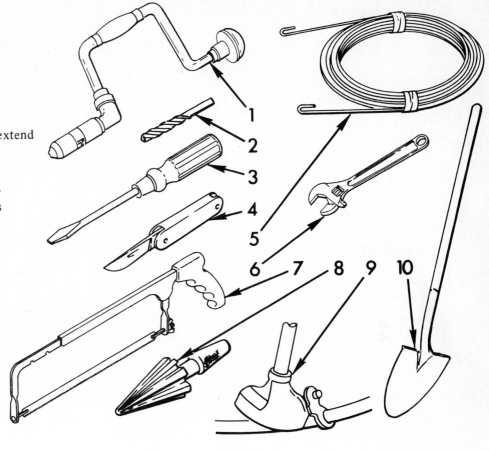

Extending Wiring Underground

The following supplies are required:

Flexible cable [1]. Purchase the type required by local code for underground installation in your area.
Thin-wall conduit [2], normally available in 10-foot lengths
Conduit ell fitting [3]
Conduit clamps [4]. Purchase the number and type required to securely attach conduit to outside wall.
Two connectors [5], for attaching conduit to ell fitting [3] and new outdoor fixture
Couplings [6]. Purchase the number required to connect lengths of conduit together.

Be sure all tools and supplies that you buy are for use with the size conduit you are working with.

Read through entire procedure before beginning.

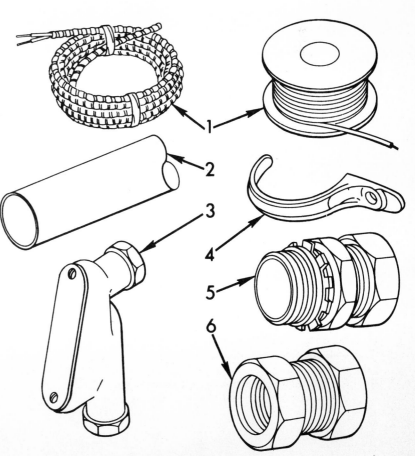

INSTALLING OUTDOOR FIXTURES AND WIRING

Extending Wiring Underground

WARNING

Be sure to turn off circuit breaker or remove fuse that controls circuit being worked on.

1. Perform procedures on Pages 77 and 78 to extend wiring from indoor power source [5] to opening [7] in outside wall. Make opening so that ell fitting [3] fits snugly.

Wires [4] from power source [5] are connected within ell fitting [3] to underground wiring. Cut cable [6] and remove insulation to allow for connecting cable to underground wiring.

2. Cut cable [6] to required length.

3. Remove cover [1] and gasket [2] from ell fitting [3]. Connect cable [6] to fitting.

Trench for underground wiring and conduit must be at least 18 inches deep. Wiring must be below the frostline for your area and deep enough to be protected from any spading or digging.

4. Dig trench from directly below ell fitting [3] to location of new outdoor fixture.

Extending Wiring Underground

Conduit [1] must be bent as required to fit firmly in trench and attach to ell fitting [2] and new fixture. Cutting may be required. Page 81 describes cutting and bending conduit.

5. Lay out conduit [1] for entire length of installation. Cut and bend conduit as required.

6. Remove locknut [3] from connector [4]. Locknut is not required for installing connector in ell fitting [2].

7. Install connector [4] in ell fitting [2]. Tighten locknut [5] against fitting.

8. Insert conduit [7] into connector [4]. Secure conduit by tightening nut [6].

9. Assemble all lengths of conduit [1], using couplings [9] between each length. Secure couplings by tightening nuts [8].

10. For maximum protection against moisture and corrosion, wrap pipe with pipe wrapping tape.

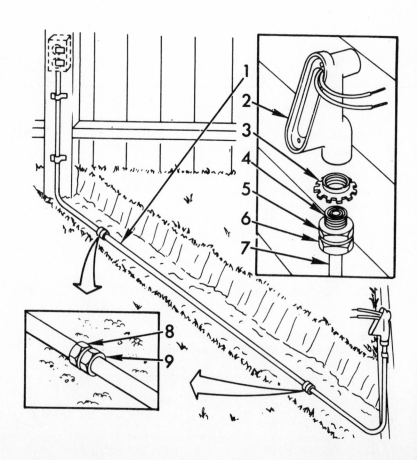

Extending Wiring Underground

11. Remove 6 inches of outer insulation from end of cable [3].

12. Fish cable [3] through entire length of conduit [8].

Local codes should be checked for the method of connecting wires [2].

13. Connecting wires [2] inside ell fitting [1]. Install gasket [5] and cover [4].

14. Secure conduit to outside wall with clamps [6] as required. Nearest clamp to ell fitting [1] should be within 3 feet for proper support.

15. Apply caulking compound around ell fitting [1] where it meets the wall.

16. Install connector [9] over cable and onto conduit [11]. Tighten nut [10].

You are now ready to install your new fixture. Locknut [7] is used to attach an outdoor electrical box to the conduit. Be sure to keep power off until fixture is completely installed. Then check the operation of the fixture and fill in the trench.

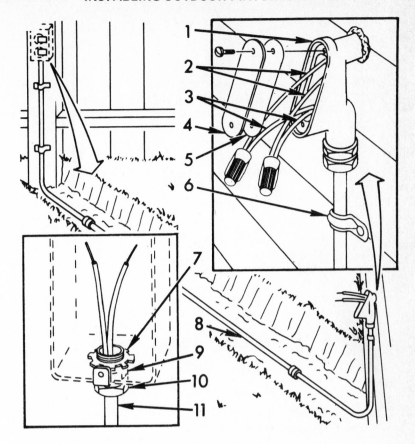

▶ Working with Thin Wall Conduit

Cutting Conduit:

Conduit is cut to desired lengths using a hacksaw [1]. A miter box [2] is helpful to insure an absolutely straight cut.

After cutting the conduit, use a reaming tool [3] to remove burrs and sharp edges which may damage wiring.

Bending Conduit:

Conduit [7] is bent using a conduit bender [6]. Depending on the desired angle [5], the size of conduit and the desired finished height [4], the conduit bender must be placed at specific distances from end [4] of conduit.

Follow manufacturer's instructions for specific distance requirements. Bend conduit [7] as follows:

1. Place bender [6] over conduit [7] at desired distance from end [4].

2. Place one foot on conduit [7]. While pressing down on bender [6] with other foot, pull handle until desired bend is made.

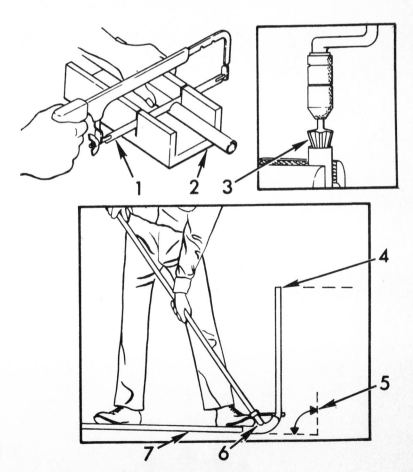

The easiest and least expensive method of extending wiring or adding outlets or switches to your home is by the use of surface wiring. This method eliminates drilling holes, fishing wires and cutting into walls and ceilings.

As with any other wiring addition, plan what you need, see what is available and check to be sure the additional power load can be handled by existing wiring.

There are several different combinations of fixtures [1] for surface wiring — usually a combination of bulb receptacles, outlets and switches.

Installation of two types of surface wiring are covered in this section:

- Plug-in Strip [2], see section below
- Metal Raceway [3], Page 84.

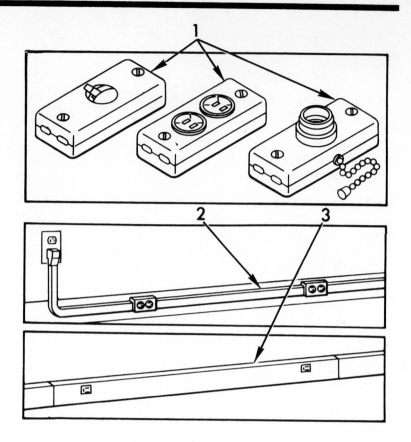

▶ Installing Plug-in Strip

Plug-in strips [3] are rigid plastic strips that can be attached to the wall or baseboard. This type is easy to install as it requires no removal of wall or fixtures.

There are two types of surface-mounted devices used with plug-in strips. One type [4] requires screws to attach the wires of the strip. The other type [5] provides wire connections by simply pushing wires into the device. Both types can be installed in as many locations along the strip as needed.

Plug-in strips [3] and devices [4, 5] are commonly produced in two or three different colors. However, you may want to paint them to match the colors of your room before installation.

The following tools and supplies are required to install a plug-in strip:

    Hammer [1]
    Screwdriver [2]
    Plug-in strip [3] and devices [4 or 5]

Read through entire procedure before beginning.

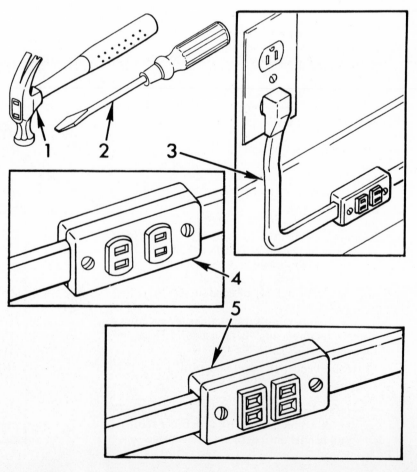

Installing Plug-in Strip

1. Connect wires at end of plug-in strip [3] to attachment plug [2]. The instructions following Step 7 describe how to connect wires to devices.

WARNING

Be sure to turn off circuit breaker or remove fuse that controls outlet [1].

2. Begin at existing outlet [1]. Plug in attachment plug [2].

3. Extend strip [3] to location of first device [4].

4. Connect wires of strip [3] to device [4].

5. Attach strip [3] to wall according to manufacturer's instructions.

6. Repeat Steps 3 through 5 to install remaining strips and devices [5].

7. Turn circuit breaker to ON or install fuse. Check operation of all devices in plug-in strip.

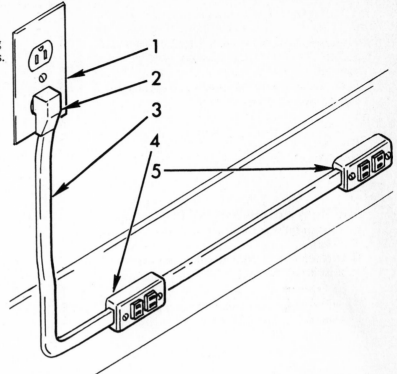

Installing Plug-in Strip

Instructions below describe connecting wires to plug-in strip devices [1] without screw terminals, and connecting wires to devices [4] with screw terminals.

Connecting Wires to Device [1]:

1. Trim off insulation [2] from wires according to strip gauge on device [1].

2. Attach wires by pushing them into device [1].

Connecting Wires to Device [4]:

1. Remove screws [3] and nuts. Separate halves of device [4].

2. Loosen screws [5]. Trim off about 1/2 inch of insulation [6] from wires.

3. Connect wires to screws [5]. Tighten screws.

4. Place halves of device [4] together. Install screws [3] and nuts.

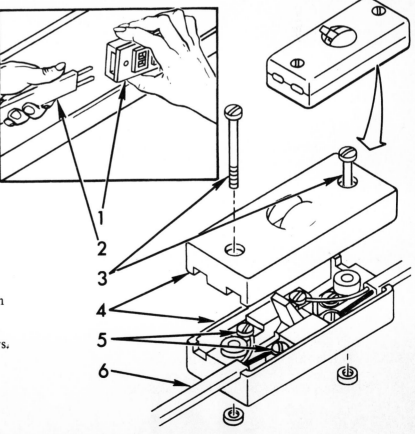

INSTALLING SURFACE WIRING

▶ Installing Metal Raceway

Metal raceways may be installed at any height along a wall or along a baseboard. If installing the raceway along a baseboard, a length of baseboard equal to the length of the raceway must be removed.

The following tools and supplies are required to install a metal raceway:

> Screwdriver [1]
> Two insulated solderless connectors [2]
> Flexible cable, the length required to extend from a nearby power source (junction box, outlet, etc.) to attachment end of metal raceway

Read through entire procedure before beginning.

1. Extend wiring [4] from your selected power source [3] to the location where raceway is to connect. Instructions beginning in the middle of Page 62 provide different methods of extending wiring.

2. Remove about 4 inches of outer insulation from cable [4]. Prepare wires [6] for connection. Page 26.

3. Connect cable [4] to inside half of raceway [5].

4. Place inside half of raceway [5] at desired position. Attach raceway to wall according to manufacturer's instructions.

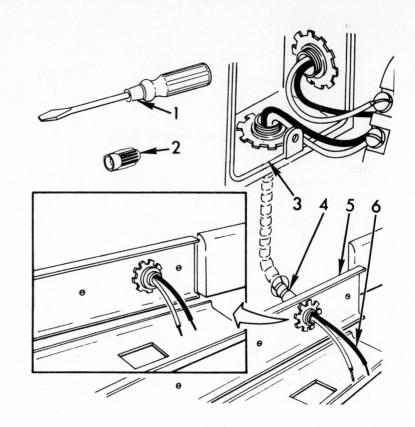

Installing Metal Raceway

5. Connect raceway wires [2] to cable [1] with connectors [3], black to black and white to white.

6. Press outlets [4] into outer half of raceway [5]. Secure outer half of raceway over inner half [6].

7. Turn on circuit breaker or install fuse.

8. Check operation of all outlets in raceway.

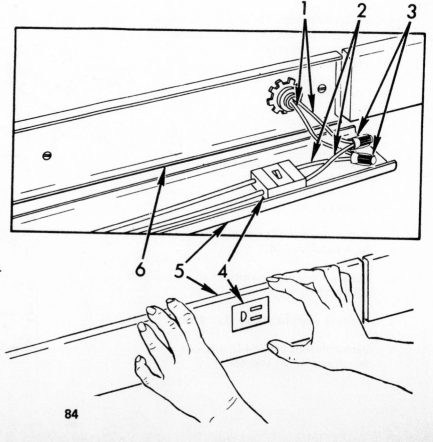

HOW TO USE A FISH TAPE

If you are extending wiring or beginning a rewiring project, the use of fish tapes will greatly simplify the job. Fishing the wires will also minimize the carpentry work required on walls and ceilings.

Fish tapes are made in different lengths. Some are on reels [1] while others are straight flexible lengths of metal [2].

The fish tape is fed through drilled holes or openings in ceilings or walls.

After holes are cut, perform the following instructions to fish wires through walls and ceilings:

1. Push fish tape [3] through hole A toward hole B.

2. While slowly turning the tape, push tape until it can be seen through hole B.

3. Grasp fish tape [3] at hole B and pull out about 2 feet. A second fish tape or length of wire may be required to grasp fish tape from behind hole B.

4. Bend hook at end of fish tape, if needed.

5. Tie cable [4] tightly to hook in end of tape at hole A or B.

6. Pull cable [4] between hole A and B by pulling tape [3] from opposite end. Disconnect cable from tape.

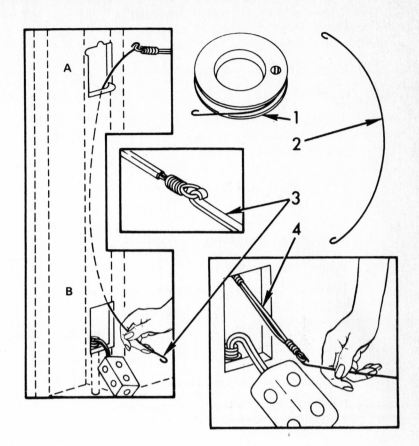

HOW TO CONNECT NONMETALLIC SHEATHED CABLE

Instructions below describe trimming nonmetallic sheathed cable and connecting it to an electrical box.

The following tools and supplies are required:

> Knife [1] or diagonal cutting pliers [2]
> Screwdriver [3]
> Nonmetallic sheathed cable connector [4]

1. Using sharp knife or diagonal cutting pliers, carefully cut and remove about 6 inches of outer insulation [6] from cable. Remove paper insulation [5].

2. Using screwdriver, punch out desired knockout in electrical box.

3. Slide connector [7] over wires and onto end of cable [6]. Tighten screws [8].

4. Install connector [7] in electrical box. Install nut [9]. Tighten nut by pushing screwdriver against slots or cogs on nut.

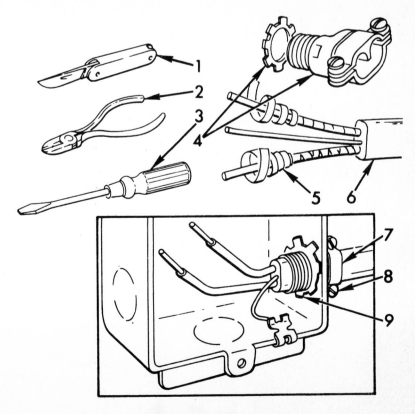

Instructions on this page describe cutting and trimming armored BX cable and connecting it to an electrical box.

The following tools and supplies are required:

Knife [1]
Hacksaw [2]
Screwdriver [3]
BX cable connector [4]
Fiber bushing [5]

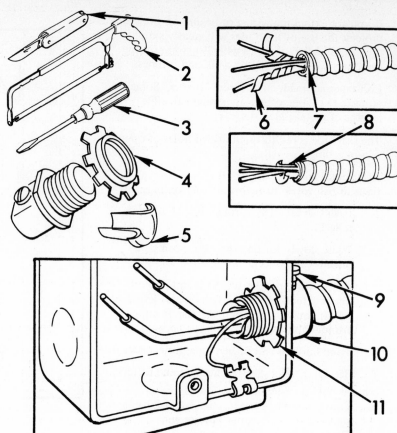

1. Using hacksaw, carefully cut through one side of armor [7]. Cut at an angle about 6 inches from end of cable.

2. Bend cable back and forth. Twist off and remove the 6 inches of armor [7]. Remove paper insulation [6].

3. Insert fiber bushing [8] into armored cable.

4. Using screwdriver, punch out desired knock-out in electrical box.

5. Slide connector [10] over end of cable. Tighten screws [9].

6. Install connector [10] in electrical box.

7. Install nut [11]. Tighten nut by pushing screwdriver against slots or cogs on nuts.

PROJECT	DATE	MATERIAL

PROJECT	DATE	MATERIAL

NOTES

NOTES

NOTES

NOTES

NOTES

NOTES

NOTES

NOTES

NOTES

NOTES

NOTES

NOTES